Developments in Environmental Modelling

Participatory Modelling for Resilient Futures

Action for Managing Our Environment from the Bottom-Up

Volume 30

Developments in Environmental Modelling

Developments in Environmental Modelling

Participatory Modelling for Resilient Futures

Action for Managing Our Environment from the Bottom-Up

Volume 30

Edited by

Richard J. Hewitt

Verónica Hernández-Jiménez

Ana Zazo-Moratalla

Blanca Ocón-Martín

Lara P. Román-Bermejo

María A. Encinas-Escribano

ELSEVIER

Elsevier
Radarweg 29, PO Box 211, 1000 AE Amsterdam, Netherlands
The Boulevard, Langford Lane, Kidlington, Oxford OX5 1GB, United Kingdom
50 Hampshire Street, 5th Floor, Cambridge, MA 02139, United States

Notices
Knowledge and best practice in this field are constantly changing. As new research and
experience broaden our understanding, changes in research methods, professional
practices, or medical treatment may become necessary.

Practitioners and researchers must always rely on their own experience and knowledge in
evaluating and using any information, methods, compounds, or experiments described
herein. In using such information or methods they should be mindful of their own safety
and the safety of others, including parties for whom they have a professional responsibility.

To the fullest extent of the law, neither the Publisher nor the authors, contributors, or
editors, assume any liability for any injury and/or damage to persons or property as a
matter of products liability, negligence or otherwise, or from any use or operation of any
methods, products, instructions, or ideas contained in the material herein.

Library of Congress Cataloging-in-Publication Data
A catalog record for this book is available from the Library of Congress

British Library Cataloguing-in-Publication Data
A catalogue record for this book is available from the British Library

ISBN: 978-0-444-63982-0
ISSN: 0167-8892

For information on all Elsevier publications visit our
website at https://www.elsevier.com/books-and-journals

Working together
to grow libraries in
developing countries
www.elsevier.com • www.bookaid.org

Publisher: Candice Janco
Acquisition Editor: Anneka Hess
Editorial Project Manager: Emily Thomson
Production Project Manager: Maria Bernard
Cover Designer: Christian Bilbow

Typeset by TNQ Books and Journals

Contents

Prologue: Some Reasons to Read This Book

Here we want to give some reasons to read the following text, which is a valuable guide containing updated tools for participatory planning.

In the first place, the focus is on "from the territory." We are not talking about a simple description of a physical or social space, but rather the territory as a kind of articulator of the different perspectives that tend to arise in administrative, professional, or academic work. Governments are divided into departments with virtually no relation at all between them, sometimes with separate areas of competence, and even vying with each other for the limelight. At the same time, professionals have been educated in distinct specialties, and in the same way tend to ignore the contribution of other kinds of knowledge. But inhabitants of the territory must understand and articulate in their daily lives all the factors and kinds of knowledge that affect it, and there are very many. The territory as an ecosystem in and of itself integrates all the energy, climate, flora, and fauna, as well as its (human) residents, who contribute their lives' experience within it. These shared natural—physical and human environment experiences require public and private policy, civil society, and the available resources to be properly integrated.

For this reason participatory planning, in the sense described in this book, seems to us to be a good answer to the problems that habitually occur in individual sectors without the necessary complementarity that any territory requires. This is not just about looking for a focus with a holistic foundation, but rather about how to apply this integrated approach in the territory and articulating the diverse techniques that can be applied in each case. It is another of the deeper reasons to apply these kinds of participatory methodologies, and the examples in this book show some very practical tools that can be applied and combined according to the diverse situations with which we are presented.

Thirdly, just a quick flick through this book reveals many different ways to use and mix the techniques described in the text. Although there is an initial diagram so the reader can distinguish one tool from another, in their practical application they always appear in combination. The requirements of each case oblige us to improvise and justify the diverse methodological articulations according to local conditions and the particular aims that we give ourselves. It is also the case that while many techniques may be applicable to the particular stages (appraisal, analysis, action, or evaluation), in reality it is not always easy to distinguish which phase you are actually in, although it certainly helps to have a preexisting logical order in mind.

In participatory planning approaches the individuals who are most involved in the territory are in charge of the process, and at the same time maintain a constant dialogue with the professionals who place themselves at the service of the participants so as to get the best out of them. It is not so much about calling as loudly as possible to see how many come to a meeting or workshop, but rather it is about understanding and listening to each sector that might have a common or divergent interest on its own terms; in this way, specific diagnoses and proposals can be developed out of the contradictions that already exist in each territory. And we hope this book will be a great help in showing what can realistically be achieved in the specific realities it abundantly describes.

Loli Hernández and Tomás R. Villasante

Acknowledgments

We would like to thank all the people who have been and are, in one way or another, involved in the various projects of the Observatory for a Culture of the Territory. Without them we would not have felt the impulse to write this book.

The writing of this book was funded in large part by the European FP7 project COMPLEX (reference 308601). We are extremely grateful for the opportunity provided by the project, which, with the encouragement and assistance of principal investigator Nick Winder and our partners across Europe, has supported us in our aim of bringing science and society closer together for a more sustainable future. In particular we would like to thank Brian D. Fath of Towson University, MD, USA and the International Institute for Applied Systems Analysis, Austria who kindly offered to include this book in the Developments in Environmental Modeling series for which he is editor. We are also grateful to the Ecology and Landscape Research Group of the Moncloa Campus of Excellence, Madrid Polytechnic University, Spain, which assisted in numerous ways throughout the process, and to Francisco Escobar of the University of Alcalá, Madrid, Spain, for supporting our approach at all times. The DUSPANAC project, which Francisco coordinated, provided us with many valuable opportunities to engage with local stakeholders in the Doñana natural protected area. In addition, we want especially to thank to all the sources of inspiration that we found along this journey, from the people of the towns, villages, and countryside who helped shape our vision to our partners in the cooperative space LaTraviesa.

Presentation

ORIGINS OF THE OBSERVATORY FOR A CULTURE OF THE TERRITORY

In the 21st century we are confronted by unprecedented changes in the territory, the majority of which are the result of land planning policies that prioritize short-term benefits over the care for natural systems that is necessary for the survival of life on Earth. In addition, on many occasions research that may allow us to appreciate the gravity of the threat does not reach the population at large, and nor is it adequately taken into account at political or administrative levels for decision-making. Politicians often seem poorly informed or unconcerned about the social, cultural, and economic problems generated by, for example, the depopulation of rural areas or the imposition of a globalized agricultural system that shows little respect for people or the environment. Citizens in general do not have easy access to reliable information about the real effects of environmental change, making the important task of refuting the myth of "growth without limits" promoted by the prevailing economic model very difficult. And though scientific work of great relevance for improved management of land and land systems is frequently undertaken, the results of this work and the implications they may have for society are often not effectively communicated to the wider public.

Against this background, the Observatory for a Culture of the Territory (OCT) was formed in 2009 and this book presents the strategies and projects we have undertaken since that time, with the aim of sharing our vision of the territory more widely. Our vision implies the construction by society itself of realistic alternatives to existing land planning models, policies, and procedures in such a way that we can begin to live within the limits imposed by the planet and share, in a socially just and environmentally sustainable way, the benefits and services that nature provides.

The principal proposition of OCT is the generation of spaces of dialogue and communication to share different visions that may exist within the same territory, inviting all relevant parties (e.g., public administrations, scientists, civil society) to participate fully on equal terms. We believe that knowledge should be available to all and should flow freely, not just from institutes and governments to citizens but also from citizens to scientists and politicians, such that decision-making is properly informed and consented to by all parties. Taking this argument to its logical conclusion implies that citizens themselves need to become active and well-informed land and resource planners. Recent events around the world serve to remind us that communities are often explicitly excluded from major territorial planning decisions that affect their interests. For example, bitter disputes broke out over the Dakota access pipeline in the United States, because obtaining the consent of indigenous communities in the locality was not thought by developers to be necessary.[1] And while indigenous communities are frequently more vulnerable because their land or water rights may not be recognized by ruling elites, violation of locals' right to decide is not something that affects only minority or first nation peoples. The recent push to open up hydraulic fracturing (fracking) in the United Kingdom resulted in ministers giving the green light to an unpopular fracking development at Fylde in Lancashire, overturning a local government decision to reject the development.[2] In both these cases the argument that the developments must take place because they are in the public interest is hard to justify given such intense local opposition. In such circumstances local communities are frequently accused of *nimbyism*[3] (not in my back yard), a patronizing accusation which carries with it three telling assumptions: the development is legitimate; the community has no right to decide the siting of the development; and a competent authority has the exclusive right to decide the site. Yet when we question the first two assumptions, i.e., if we believe that a given development is not legitimate and the community does have the right to decide its location, we also tend to question the last assumption—is the officially sanctioned competent authority really in the right? What, in any case, is this competent authority, and whose interests does it represent? Often there is great diversity of individuals or groups with a share, or a stake, in the territory, yet the interests of some of these *stakeholders* may not be represented at all in any planning procedure. And on the other hand we may sometimes suspect that some stakeholders are overrepresented—if developments that involve, for example, oil or gas are always waved through the planning system regardless of the level of popular opposition, citizens concerned about issues like climate change or access to clean water may feel that their voices are being drowned out by powerful vested interests.

In this book we contend that planning and management of the territory are the task of all stakeholders, not just these powerful groups, and moreover it is only in this way that real, shared, and sustainable alternatives to the status quo can emerge. Our concern in this book is to prepare for a future in which the citizen is not merely an interested party in territorial planning, someone whose views will be "taken

---

[1]Carasik (2016).

[2]Vaughn (2016).

[3]As Maarten Wolsink and others have convincingly argued, the accusation of nimbyism, frequently leveled against local communities protesting against unpopular developments, is often used as a strategy to discredit those who hold opposing views (Wolsink, 2006). It can be seen as an expression of frustration arising from popular rejection of an outdated mode of thinking about development siting, which regards generalized efficiency and equity concerns as the only valid criteria for choosing a particular location.

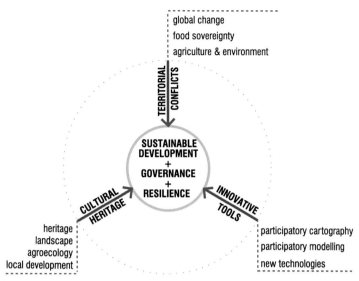

FIGURE 1

Strategic directions for participatory land planning.

into account" or, if aligned against a particular development, someone to be appeased, intimidated, or threatened with legal sanction (or worse), but an active territorial decision-maker.

To this end, the work undertaken by our group, which provides the focus of this book, has generally followed three interrelated strategic directions (Fig. 1).

- Supporting the resolution of territorial conflicts, in particular by making the decision-making process fairer and more transparent. This involves a search for consensus and negotiation between different actors at different scales.
- Development and application of innovative land management tools, particularly in rural and periurban areas, at local and regional scales.
- Building appreciation of the natural and cultural heritage of the territory through local knowledge.

About This Book: Dissemination of Experiences in Participatory Land Planning

As a concept, participatory land planning is not new, but in practice it remains unimplemented in most parts of the world. In this publication we want to contribute our experiences toward achieving this goal, through a range of projects related to the territory and its people. While the cases described are quite different, they all have a broadly participatory perspective in which the direct involvement of the different social actors provides the foundation for the creation of new models of organization and management of the territory.

To progress toward these new models, we need adequate tools that respond to the complex interaction between human beings and the territory. But since no two places on this Earth are exactly alike in terms of either social organization or physical landscape, there is no one-size-fits-all approach that can be picked off the shelf and applied to a given case. For this reason it is always helpful to undertake some background research to form a basic understanding of a given territory, its inhabitants, and its uses before launching a participatory process. This does not mean, however, that no structure or process design is possible, or that all approaches are equally likely to produce helpful outcomes. We have learned along the way, and this book is fruit of that learning process, that certain approaches to a problem work better than others, that certain techniques often give good results, and that certain habits or ways of working may prosper if sown in fertile ground. But regardless of the root causes or the peculiarities of each territory and project, the underlying approach in all the experiences compiled in this book is the empowerment of the *social actors*, the people who live and work in the place in question, to resolve the problems detected. Only in this way, through the perspective of these stakeholders, it is possible to build new approaches to the territory that are resilient enough to endure in the long term.

This book is divided into four chapters.

- The Introduction explains why we consider it so necessary to move away from existing land governance models and toward a fully participatory territorial planning, and what the implications of this perspective may be.

- The second chapter on strategies and techniques explains the methodological approach around which we have gathered the most appropriate tools[1] to deal with the processes generated in the many different case studies, which we call *experiences* (Chapter 3). To assist the reader in navigating this sometimes bewildering range of approaches, a diagram is provided (Fig. 2.2) in the form of a tree with the different techniques hanging down from its branches. A more detailed explanation of this diagram, which is central to the understanding of Chapter 2, is given in the chapter itself.
- In the third chapter we present a range of case studies showing the real-world application of the techniques explained in Chapter 2. To provide some coherence to the wide range of approaches to the various challenges faced by society in its interaction with the territory, we have divided the chapter into three subsections, each dealing with a key theme.
  - *Theme 3.1: Getting to know the territory: in dialogue with the past, understanding the present, thinking about the future* collects experiences on the appreciation and revitalization of common areas as a basis for recognition of the importance of the territory for local communities.
  - *Theme 3.2: Between city and country: building more resilient rural-urban relations* shows a range of attempts to recover the connections between countryside and city, and build new types of relationships based on mutual trust and respect between both urban and rural elements.
  - *Theme 3.3: Conflicts, citizens and society: participatory modeling for a resilient future* a range of cases explore the conflicts that occur over the uses of a territory from the point of view of the various social actors, and the options that can be generated to solve them and work toward a more integrated and participatory model of the territory.
- The fourth and final chapter presents a concise discussion in which we reflect on the lessons learnt from the experiences described in Chapter 3. The book concludes by setting out our vision of a more responsible, resilient, and democratic approach to managing the territory.

---

[1]In general we recognize the technocratic implications of the word "tool," and we have generally preferred to avoid using it where possible. However, it is difficult, and a little artificial, to avoid it completely, so we would like to clarify here that for us "tools" are discrete strategies that help us in our work, not manufactured objects that we apply to fix some mechanical system. The world is not a mechanical system, and we cannot "fix" it with tools.

Introduction: working from the ground up

1

1.1 TOWARDS A NEW MODEL OF LAND PLANNING

The impact of human activity on the biosphere is increasingly severe and represents a very serious threat to the Earth's ecosystems, which are necessary to sustain life on Earth. In the face of such impacts, which collectively we can call "global change," we cannot afford to remain impassive. Detecting and analyzing the impacts of global change and communicating this information to society have been, and continue to be, the principal tasks of environmental science, especially since a generalized environmental awareness began to take root in society in the second half of the last century (Carson, 2002 [1962]; Ehrlich, 1968; Leopold, 1949; Meadows, Meadows, Randers, & Behrens, 1972; Schumacher, 2011 [1973]).

However, while the volume of scientific studies on the problem of global change and its consequences has reached enormous proportions, and while most countries of our planet take into account, at least on paper, the need to protect the environment,[1] the unsustainable exploitation of our natural resources continues to accelerate. It is becoming increasingly clear that environmental policy as currently conceived cannot, on its own, ensure the continued heath of the planet's life-support systems on which the welfare of future generations depends.

In fact, statutory environmental protection can only be successful if it goes hand in hand with what we refer to as "a culture of the territory." A culture of the territory is more than a generalized awareness of so-called "environmental issues," because it explicitly recognizes that the place where we live, work, or spend our leisure time is the basis for all of our subsistence as well as our current and future prosperity. And while our own home, municipality, or province may be the starting point, we also need to complement this local focus with a broader understanding of its connections to the world. For example, what good is it to

---

[1]The Environmental Democracy Index provides a ranking of countries based on the strength of their environmental legislation. This does not take in account implementation. See http://www.environmentaldemocracyindex.org/rank-countries#all.

Developments in Environmental Modelling, Volume 30, ISSN 0167-8892. http://dx.doi.org/10.1016/B978-0-444-63982-0.00001-X

decarbonize the electricity systems of our own country if our largest electricity producers emit carbon elsewhere?[2]

In particular, this broader vision of the territory calls for three key things:

1. A more sustainable mode of living even where this may be non-optimal from a conventional economic point of view.
2. The abandonment of conventional economic development models that treat natural capital as if it were income (Schumacher, 2011 [1973]), and whose sole focus is the accumulation of monetary capital and the exchange of consumer goods.
3. The replacement of these models by others linked to the idea of *buen vivir*,[3] a concept that emerged in Latin America and can be loosely translated as "harmonious living" or connectedness between human communities and the natural world.

As agents who want to be part of this change, to achieve these three objectives we need to rethink the way we go about our work. A more resilient planet, with better governance and greater well-being for human and non-human populations, cannot be attained through the imposition of hierarchical top-down programs and structures, but by advocating a change of perspective in which the territory and its inhabitants come first, and are fully enabled to decide their own needs. It is not enough for us just to keep working on scientific projects whose contribution to society is measured only in academic publications or policy briefings. It is necessary to work with local communities to co-develop and increase knowledge about the territory and its governance and look for new ways to help communities take back control of their own lands. Conventional knowledge pathways, like this publication, can help steer us toward this goal, but they are no substitute for ideas and solutions that emerge from the communities themselves. To bring these processes to life requires careful identification of (in social science jargon) "stakeholders," typically those who live, work, or spend leisure time in the territory, e.g., local inhabitants, farmers, environmentalists, politicians, entrepreneurs, tourists, researchers, or anyone with an interest or involvement in a place at a given moment in time. As social researchers, finding out how to involve stakeholders in a way that minimizes discord and leads to consensual, constructive approaches to difficult problems is our most important task.

---

[2]The Swedish state-owned energy company Vattenfall is a case in point. Though its decision to transition to clean energy should be commended, the sale of its coal-mining assets in Germany to Czech company EPH was a disappointment to many who had hoped the operations would be decommissioned. While the Swedish state looks cleaner, carbon continues to be emitted from the same sources. See, e.g., http://uk.reuters.com/article/us-vattenfall-germany-lignite-idUKKCN0XF1DV.

[3]E.g., Gudynas (2011). Gudynas notes that the concept of *buen vivir* has been formally incorporated into the constitutions of both Bolivia and Ecuador.

1.2 PARTICIPATION: THE IMPORTANCE OF GIVING A VOICE TO THOSE WHO KNOW ABOUT AND LIVE IN THE TERRITORY

The process of engaging and working with stakeholders in designing and developing actions and interventions in the territory we refer to as participatory planning (Gómez Orea, 2001; Healey, 1994; Tress & Tress, 2003).

Participatory planning offers a number of advantages over conventional land planning approaches.

1. Transfer and exchange of knowledge between sectors. Each actor plays a role and provides a different bias or perspective that enriches the process and facilitates conflict resolution. This does not involve the rejection or replacement of scientific evidence, but the transfer of scientific information and approaches to the reality of civil society and vice versa. The aim is to open up the decision-making process so that realistic alternatives to current modes of development can be proposed.[4]
2. Greater involvement of society in planning processes at various levels of action (at the scale of the farm, municipality, region, etc.) has repercussions globally. This is important, since the goods and services provided by social−ecological systems[5] transcend administrative boundaries.
3. Promotion of a system of governance[6] in which all stakeholders have an important role in decision-making. This implies greater individual and community responsibility in the management of the territory, in return for greater local sovereignty over local resources.
4. Greater diversity and compatibility in the way that land is used, as a result of the need to fairly accommodate the many different visions that stakeholders provide.

Of course, compared to conventional, non-participatory decision-making processes, a participatory approach takes time, often longer than the typical 4- or 5-year period that policy-makers tend to favor in accordance with electoral cycles. Nonetheless, by securing the wider participation of all the groups and sectors involved in the territory, policy-makers are likely to be rewarded by greater acceptance and more durable policy actions. The more stakeholders "buy in" to the process, the more successful the results are likely to be. At the same time, the more

---

[4]By "decision" we refer not only the decisions that may be taken by the competent authorities at any given time, but the set of all possible decisions in the territory or "decision space" (Oxley, Jeffrey, & Lemon, 2002).
[5]By "social−ecological system" we refer to the whole system, comprising the natural ecosystems in a particular territory and the human social actors that interact with them.
[6]In this publication we equate the word "governance" with a new approach to the politics of land planning. Governance, for us, implies an inclusive, bottom-up approach to decision-making based on knowledge and consensus between all parties.

unpopular a development is likely to be, the broader the participatory decision-making process needs to be. Unfortunately, the opposite is quite usual, and developments that are likely to prove unpopular are rushed through so as not to attract attention. Typically national planning regulations include provision for "statutory undertakers," usually services or utilities providers, which are allowed to circumvent the usual planning regulations—and governments take advantage of this in cases like pipeline infrastructure. But what happens when a sitting government includes representatives of the same corporations which stand to benefit from these installations? Are such developers really always acting in the public interest? Is there any justification for providing license to circumvent planning procedures to *any* group? In general, we would suggest there is not. Developments that are alleged to be of public benefit should be decided through transparent public processes, not in private meetings.

1.3 WORKING FOR CHANGE: STRATEGIES AND PERSPECTIVES

To achieve a greater integration of all stakeholders such that communities begin to own the processes of change that are taking place in their territory requires new ways of working. Clearly, these new approaches do not arise out of nowhere; in fact a range of existing concepts and theoretical approaches already show us the way. As a guide, the principal concepts on which our work is based are briefly outlined in this section.

1.3.1 SUSTAINABLE DEVELOPMENT, THE ECOSYSTEM APPROACH AND RESILIENCE

The recognition that the planet's resources are limited implies the rejection of a development model that places no limits on the exploitation of these resources. In fact, even the word "resources" itself, as applied to naturally occurring products or systems, confers the connotation that human exploitation is inevitable and automatic. The now well-known concept of "sustainable development" emerged in the 1980s as a response to the paradigm of growth without limits.[7] However, there are some problems with the concept—in particular, its ambiguity gives rise to interpretations far removed from the essence of the concept of moderate and balanced use of natural resources and the environment that allows future generations to thrive, e.g., arguments that development is necessary to "sustain" the economy and

---

[7]The report "Our Comon Future," written by a team led by Norwegian politician Gro Harlem Brundtland (Brundtland, 1987), represented the first time that the environment and human development were discussed as part of a joint political agenda. This led to the United Nations World Environment Summit (Earth Summit) in Rio de Janeiro, Brazil, in 1992.

therefore economic criteria should take precedence over environmental or social considerations (Skippers & Nicholson, 2011). However, even if we do not necessarily accept the original tripartite definition of the Brundtland Report,[8] subsequent work on the concept has sought to redefine the relationship between the "three pillars."[9] More recently, some researchers have introduced the idea of sustainable degrowth (Gudynas, 2011; Martínez-Alier, Pascual, Vivien, & Zaccai, 2010).

Even so, sustainable development, however we choose to frame it, carries with it some implicit assumptions about the need for continuous modification of the natural environment to achieve socioeconomic objectives that are often narrowly defined. Development, even if we understand that it does not always have to involve cement, is still a loaded term. To escape from this "command and control" mentality (Holling & Meffe, 1996), which has prevailed since humans first sought to change their environment, it is clear that other perspectives must be sought. One such perspective is the "ecosystem approach," and more recently the idea of "ecosystem services." Although the term was coined in the 1970s it has acquired popularity only quite recently. A number of similar definitions exist: Daily (1997), for example, defines ecosystem services as "the conditions and processes through which natural ecosystems sustain and fulfil human life." Use of this term has become widespread, to the point where some governments have begun to incorporate it into policy. However, Waylen et al. (2014) cautioned that ecosystem services alone is not a substitute for the ecosystem approach, a broader, more integrated perspective for ecosystem management adopted by the Convention on Biological Diversity in 2000. Also, the term "ecosystem services" has attracted some criticism, in particular that the idea of services is rather anthropocentric—if sustainable development brings assumptions about the planet as a kind of clay for humankind to mold or fashion at will, the idea of ecosystem services risks the implication that naturally occurring phenomena like, for example, pollinating insects or the water cycle exist solely for our benefit. However, despite these criticisms the concept of ecosystem services, unlike sustainable development, does at least allow us to move from generic, and often politicized, debate and identify concrete actions in the territory. By physically identifying ecosystem services we can move to try to ensure their well-being. It is a framework that allows us, as Cartledge and colleagues noted (Cartledge, Dürwächter, Hernandez-Jimenez, & Winder, 2009), "to stop worrying about generalities and focus instead on impacts, mitigation, compensation, and accommodation."

---

[8]The Scottish Executive (2006) notes that "Despite their frustrations with the woolly thinking of sustainable development, many Western academics, policy-makers and practitioners have been prepared to work within the framework of its overarching guiding principles because they approve of their moral and practical intentions."

[9]In particular, O'Riordan (1998) observed that the economy is a function of society, which is itself dependent on the environment, and thus the original Venn diagram approach should be superseded by a nested "Russian doll" or sequence of decreasing circles, each contained within the other. The key point here is that it makes little sense to treat the economy, a human invention, as equivalent to the environment, on which all life on Earth completely depends.

One further concept is worthy of attention here. "Resilience" can be defined as "the capacity of a system to absorb disturbance and reorganize while undergoing change so as to retain essentially the same function, structure, identity, and feedbacks" (Walker, Holling, Carpenter, & Kinzig, 2004). The resilience perspective allows us to see the interaction between human societies and their environment as an integrated, interconnected "social—ecological" system (Berkes & Folke, 1998; Berkes, Folke, & Colding, 2000). The value of this is that it allows us to manage human interaction with nature in an adaptive way, looking to enhance or maintain the characteristics of the underlying systems rather than partitioning up the territory into mutually exclusive blocks which have more to do with the perspective of particular actors that with territorial cohesion, e.g., urban and rural, agriculture and industry, protected areas and unprotected areas. The implications of the resilience perspective for land planning are profound and lack general practical implementation. Resilience thinking shows us that protected areas like Doñana, a natural area in southern Spain that is internationally renowned for its biodiversity, will continue to degrade unless more environmentally sensitive land management practices are adopted across the whole of the catchment of the Guadiamar River, much of which lies outside the protected area (Palomo, Martín-López, Zorrilla-Miras, del Amo, & Montes, 2014). To do this requires a much deeper process of social mediation and negotiation than putting up a fence and restricting activities of particular types. All stakeholders in the territory in question must be involved in collective administration of a common resource.[10] The tools and approaches we discuss in this book show how to go about such a process in practical terms.

For this type of approach to be successful it needs to be embedded in local knowledge and the tracks this leaves in the past. By binding future development to the store of traditional knowledge of a locality held in a community's history, endogenous alternatives (Guzmán Casado, Gonzalez de Molina, & Sevilla Guzman, 2000) to the current one-size-fits-all approach to development can emerge. These are likely to be easier to implement as well as longer lasting.

Fig. 1.1 shows two approaches to land planning. Each sector or actor involved in the territory has a different vision; however, this vision is frequently realized in functional isolation (option 1). Instead we argue that all actors should negotiate a collective management approach that allows each to thrive as far as is possible in a harmonious and integrated way (option 2). As the diagram implies, this would mean negotiating an end to very large-scale, mono-functional land uses so that marginalized stakeholders and non-human actors like plants and animals have space to exist. Though in practice land planning problems are rarely as simple as Fig. 1.1 suggests, the diagram may help us open our minds to other possible interpretations of territory. We note that land planning in the conventional sense does, in theory,

---

[10]For a deeper discussion of the problems involved in management of common resources see for example Hardin (1968) and Ostrom (1990). Elinor Ostrom's work on this subject led to her being awarded the Nobel Prize in economics in 2009.

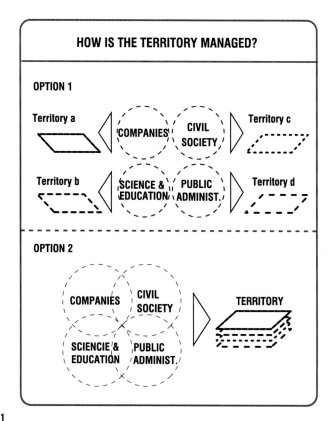

FIGURE 1.1

Two alternative models of land management.

strive to attain something like option 2. Yet negotiation between actors cannot work if all actors must submit to the crushing paradigm of endless economic growth, because the actor whose interests are best served by this paradigm will always win. Thus Fig. 1.1 reminds us that, in reality, we are far away from achieving option 2 in any real sense: most modern developments are impositions by a few powerful actors, and the idea of planning as negotiation between all actors on equal terms is frequently nothing more than colorful theatre. Yet, as we argue throughout this book, we need to try much harder to make this work. The rewards are not just a healthier natural environment, better able to sustain life on Earth, but more resilient human societies able to adapt their systems of reference to respond to changes of any sort—climatic, social, or technological (Westley et al., 2011). The current model (option 1), in which powerful actors expropriate land resources for their exclusive use and impose control over large areas without taking into account the needs of all of the inhabitants of each place, offers us only degradation, social exclusion, impoverishment, and conflict.

1.3.2 WORKING WITH DIFFERENT TYPES OF INFORMATION— INTEGRATIVE RESEARCH AND PARTICIPATORY MODELING

Working with different actors for common goals in management of land and territory is necessarily interdisciplinary and requires us to deal with both qualitative and quantitative information in a coherent way. In particular, two broad approaches, each with its own adherents and literature, are relevant: mixed-methods research (Creswell & Clark, 2007) and integrative research (Tress, Tress, & Fry, 2005a, pp. 13–26, 2005b; Winder, 2005). While these approaches are broadly comparable and share many commonalities, integrative research is better known among researchers in environmental disciplines (Macleod, Blackstock, & Haygarth, 2008). Under an integrative research approach, qualitative and quantitative information domains are viewed as linked and mutually dependent rather than separate and opposed. Integrative research approaches are especially important in the context of human–environment interaction problems. Through a process of iterative research cycles, quantitative information from measurement-orientated monitoring can be combined with qualitative approaches based on approximate or estimated data and social research.[11] This approach is illustrated in Fig. 1.2 (Hernández-Jiménez & Winder 2009).

Within the integrative research cycle (Fig. 1.2), a very broad range of tools and approaches exist which may be useful to us at particular times. We can view these as belonging to two broad strands.

FIGURE 1.2

The integrative research knowledge construction cycle.

[11]For example, Lemon, Seaton, & Park (1994), as part of their work in Argolid, Greece, experienced great difficulty in obtaining information about subsurface water levels from official sources, since locals were unwilling to disclose their water use and the location of their wells to state engineers. To obtain this information, participatory researchers worked to gain the confidence of the local population through local contacts. In this way a large quantity of qualitative information was obtained that allowed the historical evolution of the subsurface water supply to be understood, which subsequently permitted the targeted collection of quantitative data through conventional methods.

1. In the first strand we find a wide range of participatory methods drawn from disciplines as diverse as sociology, rural development, health, business management, and education. Participatory and social research tends to be thought of as purely qualitative, but in reality often incorporates quantitative numerical methods like multivariate statistics.

2. In the second strand we see that geographical and environment disciplines like agronomy, ecology, and geography offer a collection of analytical techniques with specific application to the territory, such as spatial multi-criteria analysis, land use models, and other methods from the realm of geographical information science. Geographical or territorial analyses are often thought of as quantitative in nature, but in fact frequently involve highly qualitative elements like scenario planning or ecosystem approaches.

The integration of these two strands in actions or processes to tackle human environment problems is what we refer to in this book as "participatory modeling." We return to this in more detail in Chapter 2.

1.3.3 PARTICIPATORY RESEARCH TECHNIQUES

Participatory research techniques are used to structure and facilitate the process of engagement of different stakeholders, e.g., associations and community groups, political representatives, or public administrations, with regard to a particular process or goal. Since participatory methods are found in such a broad range of disciplines, it is not surprising that a variety of different approaches exist. In our own work we have tended to favor participatory action research (McIntyre, 2007; Villasante, Montañes, & Marti, 2000). Participatory action research has three key strengths: its emphasis on breaking down barriers between facilitators and participants; the emphasis on non-hierarchical and egalitarian approaches to decision-making; and an explicit political dimension that seeks to address power imbalances within the stakeholder community. However, it is important to note that we regard this approach as a loose operating strategy for structuring information, not as a straitjacket. We recognize that other participatory approaches may be equally appropriate for knowledge co-construction with stakeholders,[12] depending on the context and objectives of the case in question.

These approaches help us to bring those actors who have remained at the margin back into the territorial decision-making process, and to allow their voices to be heard alongside politicians, administrators, and scientists who have tended to dominate the process in the past. To allow these diverse groups to build knowledge together, it is essential to generate a receptive climate that allows this process to prosper. Equally important is the careful choice of stakeholders to ensure that

---

[12]We can find, for example, *la intervención comunitaria* (community intervention) (Marchioni, 1994), social enquiry (e.g., Lemon et al., 1994), and companion modeling (e.g., Barreteau et al., 2003).

particular interests are not over or under-represented, and that scientific or technical knowledge is not prioritized over local, traditional, or informal types of knowledge. Such a framework allows us to begin to develop more inclusive governance strategies in which civil society, not technical specialists, big business, or political interests, takes the lead (Farinós, Romero, & Salom, 2009; Romero & Farinós, 2011).

1.3.4 GOOD GOVERNANCE AND *BUEN VIVIR*

Good governance, in the sense we describe above, goes hand in hand with the concept of *buen vivir* (Gudynas, 2011), encompassing the idea of well-being in the search for systems of living that offer an alternative to the current global trajectory of increasing exhaustion of natural resources and growing inequality. A range of emancipatory movements already exist that put the sustainability of life itself at the center, searching, in the words of Pérez Orosco, "for a universal framework of positive lifeways in which diversity does not signify either inequality or exclusion" (Pérez Orozco, 2014). This concept invites us to think of more intimate systems, of local-scale processes for which there is no one recipe, in which each region or country must establish its own values and indicators for quality of life (Villasante, 2015), which is a fundamental aspect of the co-construction of shared proposals. This approach allows us to reintroduce the essence of community, since it is at the local scale (that of the village or small town) that community participation and facilitation of local knowledge are most easily accomplished. But while the local scale is the most appropriate for understanding a territory's values and getting people involved in its management, it is also necessary to work at larger scales to be able to defend these values, since the most serious threats often come from higher up the chain, e.g., from the region, the nation-state, the European Union, etc. It makes no sense to remain immersed in our own village if decisions taken at higher levels of government have negative consequences for all the villages in the region—"although small is often beautiful, large is sometimes necessary" (Winder, 2007). The challenge, then, is to build on a common base established at the local level, and then work outwards to weave an extensive network of "sovereign citizens" that forces change at higher levels, unleashing a social transformation.

Strategies and techniques: a living, changing process

2

2.1 CHOOSING THE RIGHT METHODS: SCHEME OF WORK

The goals of particular participatory processes in the territory range from the concrete, like sustainable management of a community forest or stakeholder-driven restoration of a degraded wetland, to the highly general, such as stemming the tide of depopulation in a rural area or understanding the implementation of renewable energy. For this reason, a wide range of tools and approaches may be appropriate for different processes and at different points within a single process. In this chapter we address this need by presenting and describing a range of tools and approaches to help realize the goal of more open, democratic, and socially coherent land and resource planning. We do not pretend to be comprehensive,[1] and neither do we claim that every technique is appropriate in every situation. That is not the aim of this book. Instead we seek to elucidate the tools and approaches which we have ourselves found to be most useful in carrying out our work with stakeholders in the territory. Readers who find the descriptions in this chapter overly dry, abstract, or difficult to grasp outside of their context might prefer to pass directly to Chapter 3 on "experiences," where we show the application of the methods presented in this chapter to real-world cases. You can always come back to this chapter for detailed explanation of the methodological framework in which the case study work belongs at a later point. On the other hand, if you are interested particularly in the techniques of participatory research themselves or are looking to understand where, how, and when to employ them, we recommend that you read this chapter first.

2.1.1 GENERAL GUIDING PRINCIPLES FOR WORKING WITH PARTICIPATORY PROCESSES

Almost all the techniques presented in this publication imply in some way the active participation of different stakeholders. Although there are no specific rules to follow

---

[1]There are many excellent books and articles containing detailed descriptions of participatory techniques and approaches. See especially Geilfus (2008).

Developments in Environmental Modelling, Volume 30, ISSN 0167-8892. http://dx.doi.org/10.1016/B978-0-444-63982-0.00002-1

in the application of participatory processes, some recommendations are presented in the following subsections which are considered to be important to achieve successful results.

1. Participatory processes should strive to be *balanced*, in the sense of not giving too much power, prominence, or weight to one particular social group or stakeholder over others, even if that group has usually assumed the role of protagonist in the past or is accustomed to leading participatory processes. Public institutions, for example, often find themselves in the role of host or leader of participatory processes. Yet while the active involvement of these stakeholders lends authority to a process and is often very important in facilitating positive changes, we should remember that public institutions are rarely neutral, and stakeholders representing these organizations may often be protective of existing institutional arrangements and structures. It is both rational and understandable for public institutions to look for solutions within the context of prevailing models. Yet at the same time, it is necessary to challenge these models, or alternatives cannot emerge. Thus giving excessive weight to planners or public administrators may channel discussions toward unimaginative but politically acceptable outcomes.[2] Also, there is a risk that the participatory process may become politicized, which will undermine its inclusiveness and limit the effectiveness of its outcomes.

2. Participatory processes should be *representative*, to generate true sharing of knowledge and allow stakeholders to challenge each other's beliefs. If participatory processes involve stakeholders from only one group or sector (e.g., environmentalists), it will be harder to get other sectors (e.g., policy-makers, industry representatives) around the table. This is completely counterproductive—without the opportunity to speak freely to each other, stakeholders on opposing sides of a question, e.g., farmers and natural protected area managers, local residents and development companies, cannot learn from each other or search for common ground. It is easy to have a "very successful" participatory process where everyone leaves contented with their existing beliefs upheld, because no one has said anything challenging or expressed a contrary opinion. On the other hand, highly antagonistic or confrontational meetings are difficult to manage and may go badly wrong. These kinds of situations are usually the result of badly managed or "last-resort" participatory engagement activities ("they're

---

[2]Oxley et al. (2002) provide a conceptual framework for working with stakeholders in developing an integrated environmental model. They distinguish between *opportunity spaces, decision spaces, and policy spaces*. "The opportunity space represents the set of all possible choices by local actors, whether perceived or not, the decision spaces describe the perceived set of choices, and the policy space reflects the extent of intended influence of related policy mechanisms" (Oxley et al., 2002, p. 31). Complex environmental management problems require us to explore the opportunity space collectively as widely as possible—a process which may be jeopardized by involving too many institutional decision-making actors too early on.

shooting at us, maybe we should invite them to a meeting"), typical of unpopular projects such as pipelines and other mega infrastructure developments like dams, roads, and airports.

3. Participatory processes should also be *timely*. It is rarely too early to begin a participatory process, but it is quite often too late. In this regard, it is important not to be too ambitious at the outset. A gentle lead-in, with telephone interviews, surveys, or informal visits to stakeholders gradually building up to more intensive face-to-face contact, may be more rewarding than a sudden immersion in highly structured workshops.

4. Look to include marginalized or unrecognized stakeholders. Who is involved in or affected by a process but does not normally get asked to express an opinion on it? Some social groups usually do not engage in these kinds of processes—in many cases these may be indigenous groups or others at the margins of conventional power structures. However, they may also be just the opposite. Large private companies are rarely present in participatory engagement processes, yet they are ever more powerful and influential. In democratic countries public authorities are relatively easy to engage, at least at lower levels, because of their mandate to represent the public and be democratically accountable. Private companies have no such mandate. They answer only to their shareholders, are increasingly free of regulatory obligations, and manage land and resources in complex and often opaque ways. Securing their active participation in these kinds of processes is essential to understanding and changing the way such activities are carried out.

5. The participatory sessions should take place in locations perceived as neutral by the participants. Spaces such as universities, environmental education centers, or city council offices may seem like ideal locations to undertake participatory activities. However, depending of the type of project, they could have certain connotations or biases which may alter the subsequent development of the processes.

2.1.2 THE FOUR STAGES OF PARTICIPATORY RESEARCH

Any project or participatory process goes through a series of phases or stages, which run from initial contact with the stakeholders to a critical evaluation of the process and its contribution to solving the problem posed. There are a series of techniques associated with each of these stages. Broadly, we can identify four key stages.

1. Problem framing.
2. Research and analysis.
3. Action.
4. Evaluation.

Although these four stages can be ordered chronologically, from initial contact to process evaluation, two important observations can be made: the techniques applied within each stage may overlap at various points during the process, and the order in which the different techniques are applied within each stage may differ depending on the characteristics of the project. The group of techniques that we have most frequently applied in our own work is shown in Fig. 2.2.

It is important to note that this scheme is not a static methodology, but a flexible and iterative one, which can be modified and adapted depending on the characteristics and needs of each process. That is because every process is unique and does not require the same number or order of techniques to be applied. At each stage, the techniques used and the order in which they are applied may vary depending on the nature of the project, the financial constraints, or the timeframe for implementation. For instance, it is often necessary to undertake some kind of appraisal or problem framing (Fig. 2.2: appraisal techniques), yet this may not give rise to any concrete action. At another juncture, detailed analysis of data may be required (Fig. 2.2: analysis techniques), giving rise to a series of recommendations for action. In other cases, a specific action will already have been decided upon, and the question is then how to implement it (Fig. 2.2: action techniques). In still others, a social

FIGURE 2.1

Different techniques used.

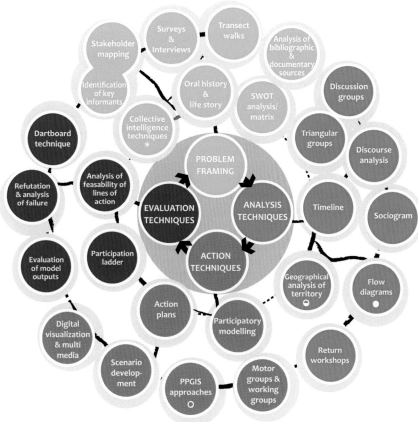

* Brain storming, snowballing.
● Problem-action matrix, problem tree, objective-solution tree.
◔ Multicriteria evaluation.
○ PPGIS survey, participatory mapping.

FIGURE 2.2

General methodological framework. This figure shows each stage within a circle of a particular color inside the trunk of a tree. The techniques that are typically associated with each stage are shown in circles, like fruit hanging from the branches of this tree, with the color of the fruit corresponding to the color of the stage to which each technique belongs. In the next section these techniques are described in order by stage, beginning at the top of the diagram (12 o'clock) and proceeding clockwise around the tree.

researcher may arrive when a series of actions have already been carried out, and be required to evaluate the impact or effectiveness of the work together with the community it is supposed to benefit (Fig. 2.2: evaluation techniques). So while an idealized model of a transformative process in the territory might call for a structured approach that progresses from appraisal to analysis to action to evaluation, in the real world this is rarely the case. Systematic evaluation of participatory processes has until recently received little attention (Blackstock, Kelly, & Horsey, 2007; Jones et al., 2009). And not infrequently social researchers are engaged too late in the process and find themselves "solving the wrong problem" (Cartledge, Dürwächter, Hernandez-Jimenez, & Winder, 2009). Thus the intention of Fig. 2.2 is to communicate this iterative and adaptive nature of any research and action process in the territory.

Finally, it should be observed that particular techniques have multiple applications and in any process may be useful at various stages—to this end, in Fig. 2.2 the tools are grouped approximately in the arena in which they are most commonly used, which of course does not imply that they should not be used elsewhere.

In the following subsections we present and explain some of the techniques applied in the course of the work, with reference to Fig. 2.2.

2.1.2.1 Problem Framing (Observation, Dialogue, Diagnosis)

Problem framing is an essential stage in the development of projects to understand the context and dynamics within which the proposed activities will operate. In general, broader stakeholder involvement at this stage of the process will allow a more robust definition of the scope of the issue to be explored and where its boundaries may lie.[3] Fuller engagement of stakeholders in this process helps to understand better the complexity of the social relations in which the problem is embedded. The problem-framing process aims to deepen understanding of the territory of interest (its geographical characteristics and related socioeconomic and cultural factors), to define the precise nature of the issues to be addressed, to identify the dynamizing or obstructing factors that may facilitate or hinder its solution, and to identify the stakeholders and their roles in the process. This can normally be accomplished in three phases: observation, where we formulate our own external subjective impressions and start to develop a network of contacts within the stakeholder community; dialogue, when we begin a process of interaction with stakeholders, enabling our initial understanding to be modified; and diagnosis, with the formation of a generally agreed understanding between all stakeholders present of the nature of the problems and how the planned activities may address these problems.

To achieve these objectives different types of techniques can be applied, depending on the nature of the project. Those that are extensively used in the projects developed by Observatorio para una Cultura del Territorio are described in the following subsections.

2.1.2.1.1 Stakeholder Mapping and Identification of Key Informants

Stakeholder mapping is a technique for rapid identification of the key stakeholders involved in a process. This provides an initial snapshot of the stakeholder

---

[3]Winder (2005) notes that "The reality judgments that characterize a model system also determine the focus of research, and can create social exclusion by putting some stakeholders 'beyond the pale'" (Midgley, Munlo, & Brown, 1998). Where projects have potential impacts on livelihoods and individual well-being, a continuing process of appreciation may be required to ensure that boundary judgments are both scientifically and ethically defensible.

community which can subsequently be further developed using the sociogram technique (discussed later).

The selection of initial stakeholders is important, since it is through them that we gain knowledge not only about the problem but also about the stakeholder community in general and the relationships within it. Specifically, the stakeholder mapping process will aim to facilitate the identification of key informants. The key informant is a strategic facilitator who may provide technical information about a region or topic, situate and orient the research team in the field, or assist in the process of selection of other stakeholders for questionnaires, interviews, focus groups, or workshops. She or he helps the researchers to approach and understand more deeply the social reality under study. However, it should be taken into account that the information derived from these key informants is likely to be conditioned by their experience. Key informants are often self-identifying, and as such are likely to have strong opinions about the role of the activities that researchers propose. Thus even (or perhaps especially) when we find ourselves in the company of a particularly helpful or efficient key informant, we should be sure to collect as broad a range of stakeholder perspectives as possible on the issue at hand. Our key informant has opened the door for us, but to begin to understand the stakeholder community we need to cast the net wider; in this regard, the sociogram technique (analysis techniques, Section 2.1.2.2.2) is especially useful.

2.1.2.1.2 Surveys and Interviews
2.1.2.1.2.1 *Questionnaire Surveys*
The questionnaire survey is a very well-known and widely-used research technique for quickly and efficiently gathering and analyzing data from a population under study. Questionnaire surveys have been used in just about every context imaginable, e.g., health, industrial relations, construction, transport planning, rural development, and many others.

In relation to the themes of this book, a brief sample of studies in which questionnaire surveys play a major role include works on population perception and attitudes to urban green areas (Balram & Dragićević, 2005; Tyrväinen, Mäkinen, & Schipperijn, 2007), identification of land-use change associated with agricultural intensification (Medley, Okey, Barrett, Lucas, & Renwick, 1995), understanding perceptions of conservation, management, and tourism issues in national parks (Ite, 1996; Trakolis, 2001; Walpole & Goodwin, 2001), valuation of ecosystem services and environmental assets (Kontogianni, Skourtos, Langford, Bateman, & Georgiou, 2001; Martín-López, García-Llorente, Palomo, & Montes, 2011), analyzing public perception of opportunities, willingness to pay for, and public acceptance of renewable energies (Devine-Wright, 2007; Rogers, Simmons, Convery, & Weatherall, 2008; Scarpa & Willis, 2010), evaluating the performance of small and medium-sized enterprises in the food supply chain with respect to sustainability criteria (Bourlakis, Maglaras, Aktas, Gallear, & Fotopoulos, 2014), and understanding how traditional ecological knowledge is retained by rural communities (Yuan et al., 2014). This is not an exhaustive list, and there are, of course, a great many other examples.

Burgess (2001) offers a basic structure to be considered when developing a survey.

1. Define your research aims.
2. Identify the population and sample.
3. Decide how to collect replies.
4. Design your questionnaire.
5. Run a pilot survey.
6. Carry out the main survey.
7. Analyze the data.

To this list we would add:

8. Critical evaluation.

Ideally, a structured process of critical evaluation should be carried out by either the research team or a group of stakeholders. It is quite frequent for only a small percentage of those contacted actually to complete the survey; critical evaluation enables us to consider the possible reason for this. (Was the survey too long and complicated? Was it circulated at a busy time of year? How were participants chosen?) The evaluation process also allows us to consider whether we actually received the information we expected, and, if not, how we might modify the questionnaire in future.

Public participation geographical information systems (PPGIS) approaches (Sieber, 2006) are increasingly widely used for collecting spatial data through questionnaire surveys. PPGIS surveys differ from conventional questionnaire surveys in that questions are asked for which responses are given in the form of a geographical location, such as "which places in the study area are important to you?" or "in which parts of the watershed have you observed degradation due to livestock?" These approaches are described in more detail in Section 2.1.2.3.2.2 on participatory mapping.

2.1.2.1.2.2 Interviews

Interviews are one of the most important methods used in qualitative research, and are frequently combined with questionnaire surveys, e.g., as part of a preliminary approach to key informants or to develop points raised through surveys in more depth. An interview can be considered to have two main objectives: to understand another person's point of view around a key topic of interest,[4] and, more specifically, to generate a *discourse* through a formalized process of dialogue with stakeholders (Hammersley, 2014). Usually the informant understands that the purpose of the interview is the former, while the interviewing researcher may place more emphasis on the latter. Some researchers, e.g., Gee (2014), observe that an ethical dilemma may arise where the informant is unknowingly providing information for the

---

[4]According to Schultze and Avital (2011), "interviewing distinguishes itself from other research approaches by engaging participants directly in a conversation with the researcher in order to generate deeply contextual, nuanced and authentic accounts of participants' outer and inner worlds."

interviewee to create a discourse with which the informant may not be comfortable. However, simple safeguards like making interviewees aware of the difference between these perspectives, checking that the discourse generated does represent the informant's reality, or involving interviewees in the analysis phase of the discourse should be sufficient to alleviate concerns in most cases.

Interviews can be structured, unstructured, or semi-structured (Longhurst, 2003). A structured interview is directed by the interviewer, the same list of questions is asked in the same order, and there is usually no room for deviation or elaboration on the part of either interviewer or interviewee. In an unstructured interview, on the other hand, the form of the interview is not predetermined and is usually directed by the informant, such as when listening to a person telling a story. However, an intermediate form of interview, known as a semi-structured interview, is often more useful. In a semi-structured interview the interviewer asks a set of questions but a more flexible and conversational approach is maintained, and the informant is encouraged to elaborate on answers or provide new information not specifically considered by the interviewer. The benefit of this is that information can be structured to address specific points of interest and allow for subsequent discourse analysis, at the same time as allowing interviewer and informant to interact more freely. Typically, a semi-structured interview with territorial actors would have the following aims.

- To understand stakeholders' perceptions about relations within the stakeholder community.
- To identify the discourses, proposals, and strategies of the different local actors and groups related to the use of the territory studied.
- To gather qualitative and quantitative data about the territory and how it is used.

2.1.2.1.3 Collective Intelligence Techniques

It is clear that the challenges currently facing 21st century human society require approaches which our existing democratic processes and institutions are unable to provide. For researcher and activist Tom Atlee, this implies the development of a form of group intelligence that is more than the simple aggregate knowledge of many individuals.[5] While it has long been clear that collective intelligence exists, recent research has found that it is measurable empirically (Woolley, Chabris, Pentland, Hashmi, & Malone, 2010). In fact, there are many qualitative techniques that can be employed to enhance group intelligence without neglecting the diversity of participants' individual insights in the identification, hierarchical organization, or resolution of problems. In practice we find that these techniques are also excellent

---

[5]According to Atlee (2014), "we do not need *collected* intelligence. We need collective intelligence, a coherent integration of our diversity that is greater than any or all of us could generate separately, just as an orchestra is greater than the sum of its instruments."

for facilitating cooperation, breaking the ice in unfamiliar group situations, and developing an atmosphere of trust. A detailed inventory of group tasks appropriate to achieve this end is beyond the scope of this book, but further information can be found in McGrath (1984). Here we limit ourselves to just two techniques, which have provided the foundation for all our recent work (e.g., Alonso et al., 2016; Hernandez-Jimenez & Winder, 2009; Hewitt, Winder, Jiménez, Alonso, & Bermejo, 2017) and are very appropriate for the problem-framing and diagnostic stage of the project: brainstorming and snowballing.

2.1.2.1.3.1 Brainstorming

This technique has the great advantage of being simple and well known. Brainstorming involves the generation, as part of a group, of creative ideas around a specific problem. The group tries to find the potential solutions to a problem based on the ideas that its members propose spontaneously.

2.1.2.1.3.2 Snowballing

This technique consists in gradually recruiting new stakeholders who have been identified by the existing participants as being valuable for the process. In this way relevant participants who otherwise may be excluded from the process are incorporated and can contribute to the research. As more stakeholders are involved, the network expands and the "snowball" becomes larger and larger, gathering momentum as the process advances.

2.1.2.1.4 Transect Walks

In a transect walk, a researcher accompanies a local stakeholder (typically a key informant) or stakeholder group on a walk through a selected territory to gain a better understanding of the territory and the way it is perceived by the informant. Different people are likely to interpret landscapes in different ways; an archaeologist, for example, may see traces of the ancient past in field systems, agricultural buildings, or roads, while a farmer's eye may be drawn to aspects like the size and accessibility of the field parcels, cropping types and their management, or the presence of a source of water. Meanwhile, an ecologist may notice birdlife, vegetation transitions, or habitat types. Clearly, when a group of stakeholders participate, the information-sharing experience is likely to be richer. This is a very good way for stakeholders to appreciate perspectives other than their own and has the great advantage of taking place outdoors, and for this reason may appeal to stakeholders who do not like to sit down in a workshop. Information provided by the informant(s) during the walk should be recorded if at all possible, provided this does not unduly affect the dynamics of the process. While the range and types of information obtainable in this way are very large, it will typically include discussion of the distribution and current state of resources, and oral narratives associated with their use. In archaeology, a walkover survey is often employed to understand a historic landscape, while specific historic elements or monuments subject to protection may be subject to periodic condition and management inspections. A transect walk may include either or both of these types of visual appreciation, being distinct from these only in that stakeholders are explicitly included in the process. Transect walks will very often lead to identification of landscape elements previously unknown to some stakeholders, and may often provide a link to specific activities like life stories or oral history.[6] Note that while the examples given here are predominantly rural, interpretive walks are also very useful in urban contexts.

2.1.2.1.5 Oral History and Life Stories

Oral history is a technique which has been widely adopted in participatory rural appraisal (Chambers, 1994b). It is often undertaken as part of wider research into the history of a particular place, activity, or people through the voices of a place's inhabitants. Life stories are therefore closely related, but by their nature are more personal and subjective. A key informant may offer his/her own life story or facilitate contact with another individual. The telling of a life story does not necessarily require a pre-established interview guide, but information about, for example, the positive and negative aspects which affected the informant's life at particular times (individual, family, etc.) should emerge in a sequential form.[7] Life stories are generally collected from people who best know the history of their region or community, frequently elderly individuals or those with a deep knowledge of previous periods. Often, particularly with very active individuals or those whose daily routine is oriented around repeated visits to the same points in the landscape, such as shepherds or farmers of scattered holdings, life stories can be developed (or emerge naturally) through transect walks.

2.1.2.1.6 Analysis of Bibliographic and Documentary Sources of Information

While social researchers often give priority to primary sources of information, striving to collect information directly from participants through interviews, surveys, or workshops, any research project is likely to involve the analysis of a wide range of other types of documentary information at some stage of the process. This typically comprises examination of scientific literature, review of policy documents, or searches of local libraries or archives for historic maps or documents. These may include written primary sources, such as eyewitness accounts, letters, acts of parliament or policy statements, or secondary sources like journal articles, books, or analyses of existing policies. Despite the importance of direct information elicitation by engaging stakeholders in knowledge-sharing processes, which we emphasize

---

[6]The Northumberland National Park Historic Villages Atlas project (2004) offers a fine example of this process at work: "The NNPA Archaeologist organized presentations or guided walks at six of the largest villages under study. At least one member of the project team participated in these presentations/walks. It was anticipated that this would help to identify knowledgeable local informants who could be interviewed further during the site visits. This proved to be the case. A more informal process of gathering such local information was undertaken during the site visits at the smaller communities under study. This process in turn assisted in selection of suitable individuals for an associated oral history project..." (The Archaeological Practice, 2004).

[7]Chambers (1994b) notes that "topics might include such as a household history and profile, coping with a crisis, how a conflict was or was not resolved."

throughout this book, analysis of existing documentary sources remains essential. It allows us to situate the present work in its proper context and focus the problem on the basis of existing knowledge before future steps are taken. This is likely to be especially important in territories with a long history of conflict over land or resources. Literature study helps us to find new perspectives on old, entrenched problems by carefully evaluating where previous approaches have proved unsuccessful. Analysis of documentary sources, like for example old maps and photographs, can itself be carried out as part of a participatory activity, e.g., in an interview or a workshop (Chambers, 1994b).

2.1.2.1.7 Strengths, Weaknesses, Opportunities, and Threats Analysis/Matrix

Analysis of strengths, weaknesses, opportunities, and threats (SWOT analysis) emerged in the 1960s as a tool for systematic evaluation of company strategy. The well-known four-panel matrix format was developed by Weihrich in the early 1980s (Weihrich, 1982) and grew enormously in popularity in subsequent years to the point of becoming, by the late 1980s and 1990s, almost ubiquitous in corporate strategy and as a result much criticized (Piercy & Giles, 1989). Some authors have advocated abandoning it completely, at least in a corporate context (Hill & Westbrook, 1997). Nonetheless, even if the demise of SWOT as a tool for planning internal company strategy is widely reported, it remains extensively used in other contexts, for example analyzing progress in environmental policy in the mining industry (Nikolaou & Evangelinos, 2010), as part of a community participation study for municipal waste management (Srivastava, Kulshreshtha, Mohanty, Pushpangadan, & Singh, 2005), and for renewable energy planning (Terrados, Almonacid, & Hontoria, 2007). One of the great strengths of SWOT analysis is its simplicity, and provided it is not treated as a solution in itself it forms an excellent starting point for collectively analyzing a particular issue or problem. In a participatory planning context it serves to identify (and subsequently prioritize) strengths and opportunities that are present in the territory under investigation, as well as the main weaknesses and threats on which to act. In this context the information gathered can be organized on the basis of either internal circumstances of the territory, both negative (weakness) and positive (strengths), or its external circumstances, again negative (threats) or positive (opportunities). Such circumstances may constitute both "risk factors" and "success factors." Excellent examples of the application of this approach in a participatory planning context are given by Srivastava et al. (2005) and Kolinjivadi, Gamboa, Adamowski, & Kosoy (2015).

2.1.2.2 Analysis Techniques

During the analysis phase all the information previously obtained is compiled, analyzed, and systematized. Some initial feedback should be given to the participants at this stage, both to maintain their interest and to stimulate further reflection within the stakeholder community on the basis of the initial problem-framing work undertaken. Clearly, the research team needs to be open to critical feedback at this point; it is still not too late to make substantial changes in the direction of the process if stakeholders do not feel that their views have been captured by the initial activities.

The analysis phase is primarily directed toward understanding the conflicts and their causes, identifying their precise nature in more detail than in the previous phase, and building a consensus within the stakeholder group around the potential alternatives and solutions to the conflicts detected. Though many of the tools described in this section are qualitative and participatory in nature, the analysis phase is likely to be stronger if supported by appropriate quantitative measurement or analysis approaches, for example land-use and land-cover change detection, borehole measurements to obtain information about water depth, quality etc., or soil testing to identify sources of pollution.

The techniques most commonly used in the analysis phase are described in the following subsections.

2.1.2.2.1 The Timeline

Timelines are chronological representations of historical and cultural events, with special emphasis on key elements connected with the object of study, e.g., land plans or policies, constructions or interventions in the territory, demonstrations or popular movements, legal judgments, migrations, technological advances, etc. This technique is frequently used in participatory rural appraisal to bring to light significant changes in the local community that influence the events and attitudes of the present (see, e.g., Chambers, 1994b), but clearly has application to any context where we want to see how past events relate to the current situation. The timeline can be drawn on any material that comes to hand, e.g., chalk and blackboard or a continuous sheet of paper fixed to the wall. The facilitator begins the process by asking participants for information about key events or milestones in a the process of interest, for example the building of a village or community, the adoption of a new technology, or the emergence of a problem and the implementation of strategies and responses to address it. If there are many participants it may be convenient to work in smaller groups, with each group electing a spokesperson to add information to the main chart. It may be useful to highlight those events considered most important by the majority of participants, or to separate events into positive (e.g., for the community, or for a common agreed goal) and negative. If the timeline is drawn horizontally, "positive" events could be recorded above the line and "negative" events below it. The timeline can be drawn on a monthly, annual, or decadal time scale, depending on the characteristics and objectives of the activity and the wider context of the investigation.

This activity is very useful as an ice breaker to encourage participants to engage in discussion and volunteer information. Though we have located this technique inside the analysis block, it would be perfectly at home in a problem-framing context, as a springboard for follow-up group activities aimed at delving deeper into some of the key events and milestones proposed on the timeline.

2.1.2.2.2 Sociogram

The sociogram is a technique of sociological and anthropological research that consists of establishing, by means of graphic representations, the social relationships existing between groups, institutions, or people. These relationships provide information about the kind of connections that exist between these groups at any given time (usually the moment at which the sociogram is developed). Importantly, it also provides a framework for formal consideration of who is and who is not a stakeholder in the context of the process we are interested in (Reed et al., 2009). Related approaches to mapping key actors or organizations as part of a participatory process can be found in Geilfus (2008, p. 40–41).

The aim of the sociogram is to achieve the following.

- Identify the key social actors at a given moment with respect to the process of interest.
- Identify the *sets of action*: following the definition given by Villasante, Montañes, and Marti (2000), we understand the sets of action as "a series of small social networks, often opposed to each other, and involved in a wide variety of processes [operating within] the conditioning socio-economic and cultural framework of each particular situation."
- Establish existing social relationships. This knowledge is a powerful aid to building or rebuilding relationships between social actors.
- Open processes about the knowledge that the different stakeholders have about their reality. The preparation of participatory sociograms with some local groups also serves to open processes of self-criticism and reflection about their own knowledge of the use and management of their territory or resources.

As we proceed through the participatory research cycle, different types of sociograms can be constructed.

The *preliminary sociogram* (Román Bermejo, 2016) aims to depict the research team's own a priori knowledge of the agents involved in the territory and their potential interrelationships. This sociogram is a good starting point for other techniques, e.g., interviews with key actors, to begin to develop a broader social knowledge of the territory.

The *individual sociogram*, carried out with particular interviewees during interviews, aims to enrich the database of identified stakeholders and their relationships with other actors and their context. It is done in a similar way to a Venn diagram, in which the group interviewed is represented in the centre of the diagram and the related actors around it. Conducting interviews with each group analyzed helps to establish the joint relationships, and a sociogram can be elaborated in a participatory manner.

The intermediate sociogram, performed in the middle of the process, aims to represent graphically all the relationships established through the sociograms previously identified.

The *positioning sociogram*, which is typically performed at the end of a participatory process, is fed by the discourse obtained previously, e.g., from stakeholder workshops. In this sociogram two axes are drawn, showing different *levels of affinity* to the proposed process objective on one axis and *degree of power* within the process on the other. Actors can then be positioned on this chart with respect to their degree of power and affinity within the process. The positioning sociogram is a useful analytical tool to apply once we have successfully collected all the relevant discourses in the territory in question within the context of the project or process.

2.1.2.2.3 Discussion Groups

The discussion group is a qualitative research technique in which a group of people (typically between 6 and 10) meet to share and discuss their views on a series of topics, e.g., those that have emerged through brainstorming or SWOT analysis; if no previous process has been carried out, the topics might be those proposed directly by the researchers[8] or by participants themselves. Through the discussion process nuances related to the proposed topics are likely to emerge, as well as the set of value judgments around the initial points raised. For example, the question "What shall we do about carbon?" carries with it the implicit assumption that "we" can, in fact, do *something* about carbon.[9]

A discussion group is thus useful for understanding group dynamics around a given issue, for example the particular views and positions adopted by its members with respect to the discussion topics posed, and how these may relate to other groups in the same process, other stakeholders, or wider society.

2.1.2.2.4 Triangular Groups

The use of *triangular groups* in social research seems to be little described, and although it is fairly well known in a Spanish-speaking context, reflections on the method itself seem to be relatively few. No examples could be found in the English research literature, and our description of this useful methodological approach is therefore derived from our own experiences, supported by the few references that describe the methodology in detail in Spanish. For Spanish speakers, a useful review of the approach is given by Ruíz Ruíz (2012), who traces the development of the method to the work of Fernando Conde, a social researcher working in the field of drug addiction (Conde, 1993, pp. 203–230). A triangular group consists of three interlocutors, and aims to create a more interactive and productive dynamic of critical reflection than ordinary discussion groups. It is intermediate between a discussion group, typically six to eight people, and a one-on-one interview. While an

---

[8]It is perfectly acceptable for researchers to propose an initial list of questions for a group to discuss, provided we accept the caveat noted by Berg, Lune, and Lune (2004): "it is important for the action researcher to recognize that the issues to be studied are considered important by the stakeholder and are not simply of interest to researchers." It might be appropriate to allocate time for the group to include questions for discussion that researchers may not have thought of.

[9]This was the title of a series of workshops held by the EU research project Complex in January 2016 in Sigtuna, Sweden. See http://owsgip.itc.utwente.nl/projects/complex/index.php/2-uncategorised/47-stakeholder-engagement-project-film-d7-5.

interview collects the personal impressions and discourse of a single individual, and a discussion group generates a multiplicity of different perspectives and visions around a range of themes, a triangular group tends to highlight differences between individuals and the possible conflicts or disagreements between them on the basic of their individual experiences.

The justification for triangular groups is clear. Conde found that discussion groups, at least in the context of his research into drug addiction, tended to close down in the direction of stereotypical responses and prevent new ideas from emerging.[10] This is probably a reflection of the dynamic of the discussion group itself (helpful in some situations), which may tend to crystallize group views around themes or opinions that are acceptable to the whole group.

The triangular group, by its nature, does not promote the search for consensus, but rather aims to transform each participant into an active subject with an individual experience that supersedes his/her social role and consequently generates an experiential discourse that contrasts with the other two participants. The participant in a triangular group is induced to deploy the possibilities of new arguments free from the constraint imposed by the need to belong to the group. Thus the triangular group allows for more open and interactive dynamic that may be more conducive to the emergence of new ideas.

2.1.2.2.5 Flow Diagrams
Flow diagrams are widely used in participatory action research and related approaches, e.g., Kesby (2000), Narayanasamy (2009). This technique consists of ordering and grouping, by graphical means, chains of issues related to the question of study itself to understand sequences of cause and effect. Flow diagrams, according to Narayanasamy (Narayanasamy (2009)), typically have the following general aims.

- To appraise the existing situation or condition.
- To monitor an ongoing program or activity.
- To assess the impact of a project, scheme, or activity.
- To design a suitable evaluation methodology.
- To examine significant changes over time.

In its simplest form, a flow diagram comprises a circle or a box representing an issue or resource with lines radiating out from it connecting to related issues or resources. The goal is to represent the process by which a particular issue functions or changes. The issue at hand could be societal in nature, e.g., rural depopulation: connecting lines might link to the causes of the decline, like demographic change, low

---

[10]It is worth quoting Ruíz Ruíz's description, citing Conde, in full for an English-speaking readership: "The discussion group, in some contexts, for example in research about drug addiction at the beginning of the 1990s, tended to close down, to redundantly repeat already fossilized stereotypes, to deny the existence of conflicts, to exclude possible alternatives, and consequently, the possibility of new social apertures" Conde, 1993, pp. 203—230 at pp. 217—218, cited by Ruíz Ruíz (2012), authors' own translation.

employment opportunities, improved transport networks to economic centers, etc. A physical resource, e.g., *upland pastures*, might seek to understand the operation of the resource, e.g., linking to the other resources implied in, or affecting, its condition or maintenance. Thus a link might be made to the availability of land workers, affected by declining rural populations. Hence these two examples can be shown to be linked, eventually appearing in the same flow diagram but starting from different perspectives. Flow diagrams can have a multitude of objectives and take a wide range of forms (see, e.g., Narayanasamy, 2009). They are often a starting point for scoring or ranking the most important causes and effects to be addressed in subsequent activities.

In the following subsection we describe three types of flow diagrams that we have found useful in our own work.

2.1.2.2.5.1 Problem–Action Matrix

In this approach, a group constructs a flow diagram in the normal way to identify the key issues and connect them to their causes and effects. Once this information has been obtained, a double-entry table or matrix is drawn on the wall or floor. Each issue identified in the flow diagram is placed in a quadrant of the table. Along the horizontal axis three levels related to the possibility of addressing each of these issues by the group are established.

- Near space: at this level the group understand that they can take actions to provoke the necessary changes.
- Medium space: at this level the group agree that they do not to have all the capacity to influence the changes. It does not depend only on the group, but the change is possible by forming potential alliances with other groups of action.
- Far space: refers to those aspects that cannot be addressed by the group in the short or medium terms.

The vertical axis establishes different levels of action (local, regional, national), determined by the information collected.

The objective of this technique is to identify and prioritize the critical points that prevent us from taking action to address the issues under investigation. These critical points are identified by arrows that enter and exit each issue.

2.1.2.2.5.2 Problem Tree

A problem tree is a type of flow diagram widely used in participatory rural appraisal (Narayanasamy, 2009; Geilfus, 2008, p. 151). The aim of the problem tree is to identify the main problems facing a particular community or territory, clarify the causes and consequences, and reveal the links between them. Like an ordinary flow diagram, issues are drawn on a chart and linked to other issues with arrows. Unlike an ordinary flow diagram, a problem tree looks to separate the key elements of the problem into a hierarchical structure, typically trunk, roots, and branches, with the trunk representing the main problem, the roots the core or origin of the problem, and the branches the effects or consequences of the problem. In addition, the institutions or groups that stakeholders consider could be involved in the resolution of those issues can be added to the tree.

2.1.2.2.5.3 Objective/Solution Tree

The objective or solution tree is the inverse of the problem tree, where the key problem, the trunk, is converted into an objective or solution to the problem, for example "deteriorating water quality" on the problem tree becomes "improve water quality" on the objective tree (http://www.sswm.info/content/problem-tree-analysis). The root causes of the problem are converted into the means by which the problem can be solved, with the consequences of the problem now becoming the ends achieved by those means.

2.1.2.2.6 Discourse Analysis

Discourse analysis is a qualitative technique that is useful for unraveling the main structures of significance from linguistic productions of various sorts, e.g., newspaper reports, speeches, legal judgments, and interviews. In participatory research it can be usefully applied to analyze the discourse of stakeholders from interviews or interaction with other stakeholders in participatory activities. The structures of significance in this case refer to the inference that the social agents obtain from their own principles, experiences, beliefs, and reflections, and they can be extracted by applying different techniques (psychological, common sense, inquiry about contradictions). Discourse analysis is a very broad field with application in a large range of situations, for example analyzing rhetoric in political speeches (De Castella, McGarty, & Musgrove, 2009), understanding the ideology of health and illness (Lupton, 1992), investigating social attitudes to indigenous land rights (Banerjee, 2000), and media representation of community opposition to extractive industries.[11] General methodological textbooks are numerous, e.g., Gee (2014), Fairclough (2013), Coulthard and Coulthard (2014), and there are a number of papers of specific relevance to discourse analysis in environmental policy and planning, e.g., Hajer and Versteeg (2005); Feindt and Oels (2005).

Analysis of information collected from these primary sources during the research is usually carried out in two stages: transcription of the materials, and content analysis. Two types of transcription can be conducted: verbatim (generally used for individual interviews), or a summary of the content (more appropriate for meetings and informal conversations). The focus of the content analysis is a search for key issues in the verbatim or compiled text for critical review by the researcher. This

---

[11]Sayago (2015) offers an example of the kind of content that might be extracted using discourse analysis in an examination of the way the Chilean media cover the resistance of a specific community to a mining business that threatens their way of life. "The discursive representations placed into circulation by each medium can be analyzed [by looking for] the justification of conflict, the description of events, the characterization of the social actors involved, the narrator's tone, the importance given to the ecological, economic and cultural aspects of the mega-mining business, the expression of expectations regarding the short-, medium- and long-term consequences of the conflict, reference to the role of the government and the State."

could involve, for example, searching for specific keywords and counting the number of times they occur.[12] Sometimes statistical treatment of the results might be appropriate (see, e.g., Yelland, 2010). By identifying the key issues in this way it is possible to gain an understanding of the most and least relevant elements in a particular discourse, and draw conclusions about the way participants or other stakeholders see the topic at hand.

2.1.2.2.7 Geographical Analysis of the Territory

This broad range of methods and approaches involves analytical study of the territory to understand the variations and distribution of its attributes. A good starting point for understanding the role of geographical analysis is Waldo Tobler's famous *first law of geography*: "everything is related to everything else, but near things are more related than distant things" (Tobler, 1970). The first law of geography reminds us that the essence of a place is bound up with its location on planet Earth, its climate, atmosphere, geology, flora, fauna, habitats, ecosystems, and human societies. This idea is complementary to the key principles of participatory planning, which seeks to capture the intrinsic, irreplaceable value of a particular forest, village, municipality, or region from the point of view of its inhabitants by emphasizing its myriad peculiarities that allow it to be differentiated from elsewhere. At the same time, by noting that everything is fundamentally related to everything else, we remember that decisions about the territory cannot be taken in isolation—that a river, for example, has a catchment, that a species has a range and an ecological niche, and that a city may be an effective mechanism for distributing services to a large number of people but it also distributes its impacts to land, water, air, and biodiversity much more widely than a village or a farm.

Given the need for a change of paradigm in land planning, in which large state-supported institutions of the past take a back seat to more active citizen planners, it is also appropriate that large public datasets of land information should be freely available and, following the Wikipedia model, eventually community maintained. This is already a reality for much geographical data, with a wide range of community mapping initiatives, e.g., Open Street Map (https://www.openstreetmap.org/), Open Topo map (http://wiki.openstreetmap.org/wiki/OpenTopoMap), Stamen maps (http://maps.stamen.com/) etc., already following this model. Other large-scale datasets created with major investment of public resources are not under community development but are available free of charge, e.g., Corine land cover (http://land.

---

[12]Doulton and Brown (2009), in a landmark study on climate change discourse in British newspapers, offer a six-point framework for analysis of the sources studied. 1. Surface descriptors (source, author, date, page, section, word count, title of the article). 2. Basic entities recognized or constructed (understanding of climate change, authority given to sources, the role of science). 3. Assumptions about natural relationships (impacts, severity, and uncertainty of climate impacts in different parts of the world). 4. Key actors and their interests and motives. 5. Key metaphors and other rhetorical devices deployed to convince readers/listeners. 6. Normative judgments (what should be done, and by whom, to solve the issues).

copernicus.eu/pan-european/corine-land-cover). To give just one example, a citizen with access to the internet can download high-resolution satellite images of most of the world, build a land-cover map using classification software, analyze a series of land-cover maps for different points in time to build up a picture of changes, and create a website or blog to make the results available all using free data and community-developed free and open-source software.

Geographical analysis of the territory is an irreplaceable element of proper land planning, since it enables us, in as objective terms as possible, to classify and quantify what exists where in a particular territory, and to study how this may be changing. As a subdiscipline of geography, geographical analysis techniques are too varied and diverse to be fully described here. They give us, for example, accurate maps of the Earth, a varied range of classifications of particular types of natural (physical) features, an understanding of what types of crops are grown and where, and what the predominant land use is at a particular location.

Geographical analysis can be usefully undertaken as part of a participatory process, integrating objective information on environmental change obtained by comparing maps of different dates with information obtained by stakeholders through workshops or interviews. This kind of integrated participatory approach is very useful for distinguishing catastrophic change events, like flooding or discharge of contaminants into a river system, from gradual change processes (Hewitt, Van Delden, & Escobar, 2014), or for highlighting the difference in perception between different stakeholders on land-cover change (Mapedza, Wright, & Fawcett, 2003).

2.1.2.2.7.1 Multi-Criteria Evaluation

Though it is clear that a wide range of geographical analysis techniques are potentially appropriate for collaborative planning of the territory, multicriteria evaluation (MCE), which seeks to optimize a range of different requirements, usually in a spatial planning context, is especially appropriate for working with stakeholders.

MCE involves the establishment of a series of criteria, values, or indicators that are subsequently weighted according to the object of evaluation. Often the result of the MCE may take the form of a geographical information systems (GIS) suitability map where the highest-scoring areas are the locations where the greatest numbers of criteria are met (Carver, 1991). MCE applications range in scope from tightly focused, quantitative models (e.g., Cheng, Chan, & Huang, 2003) to more subjective qualitative support frameworks (Kolinjivadi et al., 2015). Typical applications of MCE include siting of waste processing facilities (Higgs, 2006), wind or solar energy installations (Cavallaro, 2009; Lee, Chen, & Kang, 2009), or new urban development (Plata-Rocha, Gómez-Delgado, & Bosque-Sendra, 2011). Participatory approaches to MCE potentially have much broader application than simply seeking "optimal" locations on the basis of geographical criteria, however, for example as part of a wider analysis of trade-offs between development alternatives like expansion of tourism and natural area protection, e.g., Brown et al. (2001), or for uncovering conflicts between actors and improving the knowledge-sharing process around a common resource like water (Salgado, Quintana, Pereira, del Moral Ituarte, & Mateos, 2009).

Participatory MCE is also useful for bringing different, potentially antagonized, stakeholder communities together for open discussion about siting options. In cases of strong disagreement between developers and communities about siting of installations, MCE approaches could be used to co-construct a participatory spatial plan incorporating the views of marginalized stakeholders. Interesting examples of stakeholder-driven MCE approaches can be found in the energy planning literature, e.g., Hobbs and Horn (1997), Kowalski, Stagl, Madlener, and Omann (2009), and Alvial-Palavicino, Garrido-Echeverría, Jiménez-Estévez, Reyes, and Palma-Behnke (2011).

2.1.2.2.8 Return Workshops

A return workshop is a participatory event held with the specific goal of communicating to participants in a process about the way in which the information they have provided is being used, and collecting their critical feedback and reflection. Typically all the social actors engaged by researchers throughout the process, e.g., planners or other administrators, land managers and owners, farmers, local business-people, and other key local participants, are invited to attend. It is often desirable to present the results to a larger group (e.g., all local inhabitants) through open invitation.

The objective of the return workshop is, first and foremost, the return of the information and the conclusions gathered in previous phases to the participants involved in such a way that the information originally provided by the participants and the researchers' interpretation of this information are clearly visible. This helps to build trust among the stakeholder group and encourage reflection. However, a return workshop can also serve as a meeting point or bridge between disparate stakeholder communities that have been engaged separately by different strands of the project, thus encouraging collaboration and rapprochement of different sectors involved. In this way it can help to strengthen channels of communication between different stakeholders who under normal circumstances may have no contact with each other.

Although we have chosen to describe this technique in the section corresponding to analysis tools, there is no particular reason why return activities cannot take place earlier or later in the process.

The structure of a return workshop is determined by the information that needs to be shared, so almost any activity that makes this is interesting, accessible, and easily comprehensible and encourage some structured process of reflection are appropriate. The outcome of a return workshop, aside from the potential strengthening of trust relationships and lines of communication, should be concrete feedback on the researchers' approach and treatment of the participatory information provided. As such, application of specific evaluation techniques (Section 2.1.2.4) might be appropriate. If the return activity is a public event at which large numbers of people are present, critical feedback could be collected through questionnaire survey or via a comments board or website.

One of the most important aspects of the return workshop is that it provides an opportunity for critical reflection on work already carried out, but without the risk that the work is too far advanced for changes to be made (e.g., modify methods, models, perspectives, data, or participants). If the possibility to make major changes to the process is not offered to participants at key points in the project cycle, there is a risk that they may become disillusioned, perhaps expressing opinions like "It doesn't matter what we say, the researchers will always just do the same thing," or "They ask for our input, but then they ignore the information we give them." Of course, it is not easy to make fundamental changes to a process halfway through, but it may be necessary. In our experience, the trust gained by showing willingness to change on the basis of critical feedback received in return workshops often outweighs the difficulty of actually doing so. To be able to do this successfully, it is important not to overcommit to a particular approach too early in the process. For this reason, making expensive investments, e.g., purchasing proprietary modeling software, right at the beginning of a project or process is not recommended. If researchers are heavily committed to a particular approach, a participatory process can rapidly turn into a "we've got a great hammer, let's try and find some nails" type of approach (Prell et al., 2007), where discourse and possible outcomes are all forced into the channels required by a method or software. Viewed from this perspective, we can rapidly see that return workshops are more of an opportunity than a risk.

2.1.2.3 Action Techniques

The main objective of the action phase is to arrive, by means of the appropriate participatory methods, at a series of practical actions in the territory that respond to the problems identified in the analysis phase. Beyond this, the action phase seeks to provide opportunities for return of information (see return workshops, above) and critical feedback from stakeholders; create spaces for interaction between different social actors by strengthening existing networks and creating new ones; facilitate critical evaluation of the process itself; and establish a clear timescale for follow-up of the proposed actions at a later date.

Putting it all into practice is of course the most difficult part. To achieve this key objective, we look to a range of techniques that favor interaction and group thinking and foster new forms of social creativity and conflict resolution. We discuss some of these in the following subsections.

2.1.2.3.1 Participatory Modeling

Participatory modeling (PM) involves the integration of *participatory methods* drawn primarily from social disciplines, like sociology, business, operations research, health, and rural development, with *analytical modeling approaches* for solving complex problems. It is increasingly widely used for territorial planning and decision-making in situations where environmental degradation is acute, uncertainty is high, and there is no widespread agreement among stakeholders about what to do. Well-known PM approaches include companion modeling (Barreteau et al., 2003), mediated modeling (Van den Belt, 2004; Antunes, Santos, & Videira, 2006), and the integrative catchment modeling approaches of Voinov and colleagues in Chesapeake Bay, e.g., Voinov & Gaddis (2008), Gaddis, Vladich, & Voinov

(2007). PM is typically useful for addressing questions which clearly may not have a single straightforward answer, such as "what must we do to conserve this wetland for future generations?," "how can we improve water quality?" or "what impact will renewable energy development have on existing land use?" Questions of this kind are known as ill-posed problems, and are common in human—environment interaction (Winder, 2003, pp. 74—90). To address them requires a range of flexible, integrative approaches that recognize their inherent complexity and effectively incorporate the social dimension upon which successful resolution of any such problem ultimately depends.[13] For example, a conventional mathematical model might be employed to understand the flow of nutrients into a river, and to model the dispersal of those nutrients in the water system. But to begin to *solve the problem* of nutrient pollution, a wider process of engagement with local communities, planners, landowners, and businesses is necessary. For example, in the Solomon's Harbor watershed, Maryland, analytical modeling tools were used to analyze nitrogen from three anthropogenic sources (septic tanks, atmospheric deposition, and fertilizer), while a participant group drawn from the local community worked to improve model assumptions and model relevance to the local policy context. The stakeholder group then collectively developed a series of recommendations for local water policy (Gaddis et al., 2007). PM need not necessarily involve numerical models—qualitative modeling frameworks may sometimes be appropriate. For example, contextual interaction theory is typically applied by researchers to understand how different actors may respond to a particular directive or policy, such as a river re-naturalization program (De Boer & Bressers, 2011). A PM approach might draw on this theoretical framework but involve policy stakeholders directly in participatory activities to evaluate their own responses to the policy in question, arriving at some plausible estimation of future outcomes (Hewitt, Winder, et al., 2017). In either case, a participatory model is an excellent way of bridging different disciplines or knowledge communities (Fig. 2.3).

The objectives of PM are thus quite diverse, but often include the following.

- Propose alternatives and generate new ideas.
- Incorporate expert knowledge held by non-traditional stakeholders (e.g., farmers, fishers, local people).
- Help resolve conflicts by bringing antagonized stakeholders together to work on a common problem.
- Improve understanding of complex issues.
- Help scientists to understand that non-scientists have valuable knowledge to share.

---

[13]As Gaddis et al. (2007) noted, "in problems characteristic of highly complex systems, when facts are uncertain, values in dispute, stakes high and decisions urgent, there is no one correct, value neutral solution. Under such circumstances, standard Western scientific activities are inadequate and must be reinforced with local knowledge and iterative participatory interactions in order to derive solutions which are well understood, politically feasible, and scientifically sound."

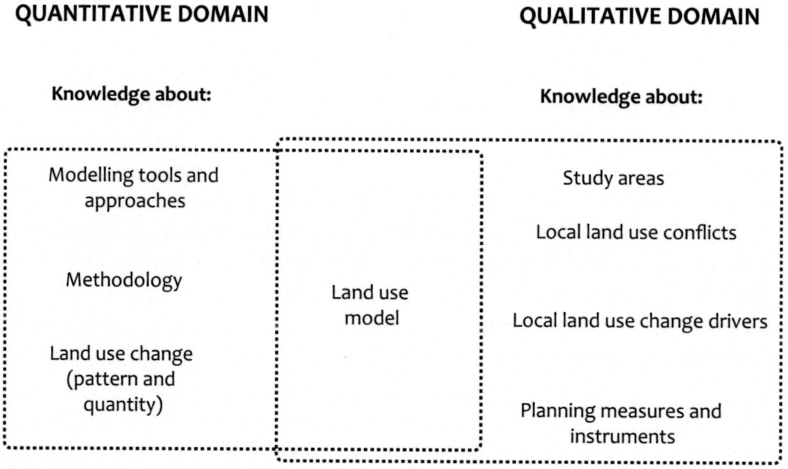

QUANTITATIVE DOMAIN   QUALITATIVE DOMAIN

Knowledge about: Knowledge about:

Modelling tools and approaches		Study areas
		Local land use conflicts
Methodology	Land use model	
		Local land use change drivers
Land use change (pattern and quantity)		Planning measures and instruments

FIGURE 2.3

A participatory land use model as an integrative approach to sharing knowledge across different domains.

After Hewitt, R., Van Delden, H., & Escobar, F. (2014). Participatory land use modelling, pathways to an integrated approach. Environmental Modelling & Software, 52, 149–165.

- Help non-scientists to understand that science does not need to be elitist, exclusive, or impossibly difficult.
- Encourage acceptance of policy support tools and approaches.
- Negotiate buy-in with local communities.
- Build trust between different knowledge communities and social groups.

PM is by nature iterative, and often involves several phases of information exchange between analytical phases (e.g., land-use change analysis, land-use suitability mapping, policy analysis) and discursive phases (e.g., problem framing, participatory model calibration, participatory evaluation of results) (Hewitt et al., 2014) to arrive at a convergence of perspectives around the issue, leading to a change of understanding among the stakeholders or "social learning" (Reed et al., 2010; De Kraker & van der Wal, 2012).

Finally, an oft-cited aim of environmental modeling studies is decision support. While it true that the precise meaning of the phrase "decision support" is context dependent, we personally prefer to emphasize the role of PM as a support framework for generating new possibilities and offering alternatives, a kind of "policy-option generator" (Oxley, Jeffrey, & Lemon, 2002) rather than a "turnkey itinerary" (Barreteau et al., 2003) for steering stakeholders toward some kind of optimum set of conditions or presupposed right answer. For this reason, techniques like MCE (see previous subsection), which tend to emphasize the search for optimal solutions, are probably more safely positioned as one element of a PM process rather than as a stand-alone tool; we view PM as a framework for communal knowledge sharing and co-construction of solutions to complex problems, rather than as a single tool or technique. Thus where the problem demands it, stakeholder interest is strong, and resources are available, we recommend PM as an overarching approach for developing bottom-up solutions, as the title of this book implies, rather than as a single element in a wider process.

PM would necessarily be initiated in the early stages of any project, and follow the whole participatory cycle from problem framing to analysis, action, and evaluation. Nevertheless, the clear focus of the PM process on development of specific actions means that it is appropriate to locate it in the action phase.

2.1.2.3.2 Public Participation Geographical Information Systems Approaches

PPGIS refers to a subdiscipline of GIS in which groups of citizens are involved in collective spatial decision-making using geospatial technologies. PPGIS therefore encompasses a diverse range of applications and contexts, and this book does not aim to describe or review all the possible uses of this approach. There are several good summaries of PPGIS which the reader is encouraged to consult (e.g., Brown & Kyttä, 2014; Obermeyer, 1998; Sieber, 2006). We limit ourselves to two approaches, PPGIS survey and participatory mapping, which we have found useful in our own work.

2.1.2.3.2.1 Public Participation Geographical Information Systems Survey

In a PPGIS survey, a group of citizens is asked to complete a questionnaire survey (see Section 2.1.2.1.2.1 above) in the usual way. The difference between an ordinary questionnaire survey and a PPGIS survey is that some or all of the question responses are given by locating GIS features (points, lines, or polygons) on a map. Participants provide the spatial information using paper maps (Brown & Pullar, 2012) or internet-based mapping tools like Maptionnaire (Fagerholm et al., 2016). Results can be analyzed using conventional statistical techniques, e.g., Analysis of Variance (Brown, de Bie, & Weber, 2015), or spatial analysis approaches such as regression analysis (Bagstad, Reed, Semmens, Sherrouse, & Troy, 2016) or cluster analysis (Brown, de Bie, & Weber, 2015).

2.1.2.3.2.2 Participatory Mapping

Participatory mapping is a PPGIS approach in which a stakeholder group works together to construct a map, either building the map directly using whatever tools are available (e.g., paper, pens, the ground, sticks, pebbles), or by using gaming tokens or counters to represent elements they wish to portray on top of an existing map. One of the great benefits of participatory mapping over other techniques involving written information is that the capacity for representing familiar territory spatially is near universal and does not depend on the participants' level of formal education (Chambers, 1994a). Recently the use of large-scale touchscreen computers or tablets has become common, so a group of participants can discuss particular options and draw them on screen as they do so (Flacke & de Boer, 2016). Whatever the methodology employed, it is useful to record the interaction that takes place during the map development process for later discussion and analysis.

PPGIS approaches can easily be used at any point in the project cycle. The two methods described here flexibly bridge analysis and action phases, as, for example, in the case of participatory mapping to identify areas where stakeholders believe change is most likely to occur under different future scenarios (Hewitt, Hernández Jiménez, Román Bermejo, & Escobar, 2017).

2.1.2.3.3 Scenario Development

The development of future scenarios can play an important role in understanding and reflecting on the consequences of current actions in the territory and exploring the possible outcomes of proposed land management alternatives. Broadly, scenarios enable us to experiment freely with tendencies identified in the present to inform and guide the decision-making process. Scenario development also offers a means of bounding different outputs from a model or process where the uncertainty is high due to the instability of the modeled system (Hewitt and Diaz-Pacheco, 2017). This is typically the case for most human–environment interaction models, as well as any quantitative model where the final model state is extremely sensitive to initial conditions (Lorenz, 1972). Thus future scenarios can be developed that try to encompass all possible outcomes, from a situation in which nothing changes (stationarity), to continued development in line with observed tendencies (business as usual), through to very extreme scenarios, e.g., total ecological collapse or a completely environmentally sustainable society. Scenarios can be both plausible and utopian. While the former may be seen as more believable by policy-makers or other key stakeholders, more idealistic scenarios provide an opportunity to "think outside the box" about less plausible but more desirable realities (Hewitt, Hernández Jiménez, et al., 2017). Transformative innovations, like universal suffrage, solar energy, satellite geo-positioning, and vaccination, once seemed extremely implausible to most people. If we cannot imagine a future that we want, we stand little chance of attaining it.

A number of standard approaches to scenario development can be found in the literature. Participatory scenario planning, particularly prominent in the socioecosystems literature, emphasizes the development of scenarios from the bottom up, out of narratives that emerge from the discourse of stakeholders (Oteros-Rozas et al., 2015). In the storyline and simulation approach (Alcamo, 2008), stakeholder-developed narratives of change are used as inputs to simulation models to project future scenarios as maps or graphs. Since no special expertise is required to imagine what the future might be like, scenario-building exercises can potentially involve anyone and everyone. Under the Prelude project, European citizens from all walks of life were involved in the development of narratives which were used to create five scenarios of land-use change for the years 2005–2035 in the Metronamica land-use model (Van Delden et al., 2005; Volkery, Ribeiro, Henrichs, & Hoogeveen, 2008); see Table 2.1.

In practical terms, scenario building can be incorporated into virtually any kind of participatory activity, but in general probably belongs toward the end of a process cycle. A participatory scenario development exercise is a good way of narrowing down the possible future outcomes in the territory of interest and developing concrete

Table 2.1 The Five Prelude Scenarios from Storylines Developed by Workshop Participants

Scenario Title	Main Drivers of Change
1. Great Escape: Europe of contrast	Globalization, climate change, decreasing solidarity, and passive government
2. Evolved Society: Europe of harmony	An energy crisis, growing environmental awareness, and active rural development
3. Clustered Networks: Europe of structure	Optimization of land use and strong spatial planning in response to an aging society and a declining agricultural sector
4. Lettuce Surprise U: Europe of innovation	Growing environmental awareness, technological innovation, and decentralization
5. Big Crisis: Europe of cohesion	Climate-change-related disasters, increasing solidarity, and strong EU policy interventions

Volkery, A. & Ribeiro, T. (2007). Prospective environmental analysis of land use development in Europe: from participatory scenarios to long-term strategies. In Amsterdam conference on the human dimensions of global environmental change "Earth system governance: Theories and strategies for sustainability".

actions to resolve key concerns about changes or tendencies that emerge from stakeholder discourse. Very colorful visual representations of the scenarios, made by hand by stakeholders using collage or drawing materials, are often very successful. It has occasionally been suggested to us that these kinds of activities involving drawing, painting, or cutting pictures from magazines might make some stakeholders feel patronized or "sent back to school." In our experience we have never found this to be the case: in general we find that most people enjoy the opportunity to express themselves creatively in this way, and these kinds of activities are often highly successful. We resist the idea that some kinds of activities are more "serious" than others, and we wonder if there are some unspoken prejudices at work here.

Sometimes professional artists may be recruited to depict the scenarios, as for example for the *Doñana eco-futuros* by Antonio Ojea (see Fig. 4 in Palomo, Martín-López, López-Santiago, & Montes, 2011); in other cases, 3D visualizations or virtual worlds might be used to bring the scenarios to life (see, e.g., Sheppard et al., 2011).

Key outcomes of participatory scenario working might include the following.

- Isolation of the most important tendencies in a territory by local stakeholders who know it best.
- Involving the real decision-makers (e.g., land planners and policy-makers) in experimentation with outcomes for which they have no time in their day-to-day roles, helping inform their future decisions.
- Stakeholder reflection and self-directed learning about the possible consequences of current tendencies or modes of behavior.

- Codevelopment by stakeholders of action plans to address these.
- Finding ways to improve management of a territory under high uncertainty.
- Encouraging a wide range of social actors (e.g., from civil society, as in Prelude) to "buy in" to the land-planning process by developing their own solutions to complex problems.
- Helping society at large to get involved in deciding on the future they want for their territory.

2.1.2.3.4 Digital Visualization and Multimedia

Digital visualization approaches vary from quite simple, such as individual three-dimensional (3D) images of architects' designs for single buildings or "bubble-world" 3D panoramic views from a particular location from multiple mosaic images, to more sophisticated "fly-through" virtual visits and, ultimately, completely immersive virtual worlds in which the user can interact with avatars and virtual objects complete with touch sensations and sound effects. These latter advances for the moment remain in the domain of research, but technology is advancing all the time. Virtual reality (VR) GIS and augmented reality GIS have been around since the 1990s, and are becoming increasingly sophisticated. The potential of these approaches is enormous. Combining, for example, the powerful virtual-world capabilities of applications like Pokémon Go with cheap and widely available VR viewers like Google Cardboard (Boulos, Lu, Guerrero, Jennett, & Steed, 2017), it would be possible to create a citizen planning tool in which participants could compete with each other to plan sustainable cities or infrastructures, or explore the effect on their avatars of environmental degradation. Scenario-planning activities can be dramatically enhanced by high-resolution 3D visualizations (Sheppard et al., 2011). The next step is to create immersive scenario experiences enabling stakeholders to walk right into the future worlds they have created. Although these approaches clearly have enormous potential, they remain extremely resource demanding (computational and in terms of project time), and it is important to ensure that such intensive VR approaches should not become a substitute for a carefully planned participatory process. Many of the techniques discussed in this book can be easily accomplished with minimal resources in areas with no internet connection or even electricity. In our technology-obsessed world, the benefits of VR approaches need to be honestly evaluated for their effectiveness alongside simpler, cruder methods before making major commitments.

2.1.2.3.5 Motor Groups and Working Groups
2.1.2.3.5.1 Motor Groups
One of the fundamental elements of action research processes is the motor group (MG). The MG is a heterogeneous group of volunteers who work on a regular basis throughout the project to facilitate its implementation. Examples of the use of MGs can be found (in Spanish) in the work of Villasante et al. (2000) and others (Sepúlveda Ruiz, Calderón-Almendros, & Torres-Moya, 2012). Although it is hard to imagine that English-speaking researchers have not made abundant use of this

technique, literature searches for "motor groups" and synonymous terms like "driver groups," "facilitator groups," and "dynamizing groups" drew a blank. Possibly these terms have become conflated with well-known terms like "discussion groups" or "focus groups." An MG, however, is a different animal, and quite particular to action research. Typically, the MG will have a stable core of participants but may grow throughout the process. Ideally the MG would be formed in the initial problem-framing or diagnostic phase of the project, incorporating new participants if necessary at any moment subsequently. This group has two key functions inside the process or project: it collects and provides information, e.g., about the current situation in the territory or existing relationships between social actors; and it is fundamental to the participatory development of the project. Ideally the MG should not comprise representatives or advocates of any particular stakeholder group, should not reflect pre-existing tensions, and should play a collaborative rather than protagonists role (Villasante et al., 2000). The role of the MG may vary, depending on the special expertise or interest of its participants, and may change as the process evolves. MGs normally play a stronger role in any given process than other kinds of project groups found outside the action research context, like advisory boards or steering groups. Clearly, however, MGs might plausibly evolve out of these existing structures under the right conditions.

2.1.2.3.5.2 Working Groups

The other group that is fundamental to the action phase is the working group (WG). The WG is a team of actors who either volunteer or are selected by the participants according to their specific abilities, interests, knowledge or competencies to fulfill a certain goal.

Unlike MGs, WGs are constituted during the action phase. For example, Román Bermejo (2016), in the case of a transition to agroecology in the Alpujarra region of Granada, Spain, describes how MG participants developed transition plans leading to formation of six WGs: WG1 on ecological fig cultivation; WG2 on organic fruit and vegetables; WG3 on organic livestock; WG4 on organic olive and olive oil production; WG5 on short-distance commercialization; and WG6 on transformation of organic fruit and vegetable production. These groups were exclusively composed of farmers and livestock keepers, and had aims as diverse as searching for support from local associations and cooperatives (WG1) to understanding how to organize local markets (WG3) and providing specific training in organic horticulture (WG5) and organic production methods (WG6). The organization of small groups in this way, each with specific tasks to undertake as part of an interdependent and mutually supportive network, greatly enhances the possibilities of enacting real change (Table 2.2).

2.1.2.3.6 Action Plans

The development of action plans is a key objective of the action research phase. Based on the priorities established in previous phases, a series of actions are proposed; and for each of these actions a series of interventions are specified to be carried out within a given period. The action plan can be structured to answer

Table 2.2 Formation of Working Groups for Agroecology in the Alpujarra de Granada Region (Román Bermejo, 2016)

WG by Sector	Date	Objective	Social Actors in This Group	Content to Be Addressed	Results	WG Created by
Olive and olive oil production	May 2008	Look for opportunities to support Flor de la Alpujarra Cooperative Society from Granada Centre for Research in Organic Agriculture and Rural Development	4 members of board of governors, 2 representatives and general manager of cooperative	Problems, potential demands	2 lines of work: training in the workings of an olive press; training in olive cultivation	Flor de la Alpujarra Cooperative Society
Figs	June 2008	Evaluate continuation of collaborative working	23 members of Contraviesa Ecológica Association	Problems, potential demands	4 lines of work: control of plagues and diseases; local varieties; training; transformation	Contraviesa Ecológica Association
Short-distance commercialization (SDC)	August 2008	Examine situation regarding organization of local markets	Las Torcas Cooperative Society and Padre Eterno Association	Working on SDC; have 2 potential groups of producers and consumers in mind in Orgiva and Cadiar; no meeting so far with any potential producer	2 lines of work: dissemination of the importance of SDC in the county; Torcas cooperative society to organize local market in Pitres Researcher will organize contact with town council and first producers' meeting.	Las Torcas Cooperative Society and Padre Eterno Association
Livestock	February 2009	Return results from 2008 work and prioritize critical points of conflict	10 livestock farmers of cattle, sheep, and goats	Development of collaborative actions	7 lines of work: associationism; livestock farming; livestock feed; transhumance; slaughter; commercialization; dissemination	10 livestock farmers of cattle, sheep, and goats
Fruit and vegetables	March 2009	Training in organic market gardening	24 producers	Early stages of organic agriculture and development of collaborative actions	3 lines of work: training; fertilizer plan; general organization	16 farmers
Transformation	October 2008	Training in preparation of organic produce	12 producers	Training in procedures, opportunities, and obstacles	5 lines of work: training; pilot projects; visits; organization; opportunities for transformation	7 producers

fundamental questions such as "What?," "For what?," "How?," "With what re-sources?," "In what time period?," and "Who should get involved?" This process is described very clearly by Geilfus (2008, p. 185). Action plans should link explic-itly to activities undertaken earlier, like the problem—action matrix or objective tree (see Section 2.1.2.2.5). Typically, action plans assign tasks to WGs.

2.1.2.4 Evaluation Techniques

While the successful implementation of actions that address, at least in part, the problems identified is an important indicator of success, in many cases we are drawn to particular problems or conflicts in the territory precisely because they appear insoluble, so we should not expect these intractable problems suddenly to give way just because we apply participatory methods. Often our approach may simply contribute a new way of looking at the problem, beginning a process of *social learning* and a gradual change of perspective which we hope may bring real change later on. On the other hand, we cannot simply throw up our hands and say "we tried our best" without offering some objective criteria which show that we have made a contribution. This requires a much more rigorous and structured approach than sim-ply asking if everyone had a good time or if they thought that the process they have been engaged in was worthwhile. Social conventions dictate that you try not to be too rude to people who ask your opinion and buy you lunch, so normally asking for feedback directly is not likely to yield useful information. Circulating anony-mous evaluation questionnaires during a workshop is not a guarantee of obtaining useful critical feedback either, since participants may feel relieved that the process is over ("let's just tick all these boxes and get out of here"), allow their opinion to be swayed by their neighbors, or find that they do not have time or energy to reflect properly at the end of the workshop. A survey they can take away and answer in their own time is likely to obtain more useful responses, but it can be a risky strategy, especially if the workshop participants are busy with their day jobs or other respon-sibilities: it is quite likely that the process will fall to the bottom of their "to-do" list once they leave the workshop. Evaluation activities need to be properly thought out and properly structured if they are to yield good results. And although the evaluation section of this chapter appears at the end (after all, you cannot evaluate something until you have done it), this does not mean that all evaluation activities have to be left till last. Actually, we caution against doing this, since it limits the opportunity to make use of the evaluation results. It is much better to develop a structured program of evaluation activities at key milestone points all the way through the process.

It is also important to stress that the evaluation process is not just about helping researchers feel they have done a good job, or providing objective assessments of funded activities to assessment bodies hungry for performance indicators. It is also a very important part of the process of making the tools and actions used avail-able to the actors involved in the process to increase their autonomy and reduce their dependence on researchers or facilitators. The evaluation phase is not only about critical feedback and assessment of the value of the process, but also a way of inten-sifying actions that strengthen the transfer of the collective leadership. Thus the community evolves the ability not just to appraise the situation (problem framing/ diagnostic stage), analyze the challenge and resources (analysis phase), come up with solutions, and deploy them (action phase), but also to evaluate critically their own progress toward agreed goals. At this point full transfer has been successfully achieved, and the withdrawal of facilitating stakeholders should not jeopardize the now-established pathways of change.

To achieve the objective of participatory evaluation, different types of techniques can be applied, some of which are highlighted below. Only those activities that relate specifically to evaluation are described, since many appropriate techniques presented in the preceding sections can simply be applied to specific evaluation objectives, e.g., discussion groups or triangular groups aimed at analyzing process outcomes, surveys or interviews for collecting critical feedback etc.

It is important to make one final observation before describing the evaluation techniques themselves, which we might call "the lunch problem." The lunch problem refers to the fact that facilitators and participants will naturally get to know each other over the course of an activity or process. A wise facilitator usually tries to make workshops worth attending by providing plentiful breaks, refreshments, and often a good lunch at the project's expense. If the work is undertaken in pleasant surroundings, the last day of activities might include a tour of a natural area or a visit to a local historic site. These pleasant and cordial occasions are absolutely necessary for maintaining the goodwill that sustains often highly demanding participatory activities, but they are not in the least conducive to frank assessment of the true value of an activity or workshop. Thus a group of stakeholders who have carried out a series of exhausting participatory activities and have mostly enjoyed themselves, when asked by facilitators whose company they have also mostly enjoyed, after a relaxing lunch or a pleasant country visit, whether they think the process was any good or not will naturally tend to say "yes." If pushed, they may be able to come up with some aspects that they feel could be improved. But their true feelings are unlikely to emerge. The answer to the lunch problem is not to make the workshops unpleasant, remove refreshments, or treat participants with hostility. Instead, we suggest the following precautions.

- Evaluate the process continuously, and do not leave all the evaluation right to the end. Tired participants who simply want to go home will just tick all the boxes to get it over and done with.
- Do not evaluate by asking questions like "did you find the process worthwhile?" or "do you think these tools are useful?" because they will not provide useful critical feedback. Of course, these questions are often necessary to ask as a part of normal social exchange, but the answers (inevitably various elaborations around the theme of "yes") do not need to be recorded.
- Try to evaluate anonymously where possible. Where this is not possible, be creative with the construction of the evaluation questions, so that it is not immediately obvious which responses reflect positively on the process (e.g., by evaluating "difficulty," see below).

- If resources are available or the context permits it, evaluation should be carried out by a different group of facilitators. This is established practice in development and cooperation circles, but not so easy to do in a research project.
- Devise roundabout ways of getting the same information. For example, instead of asking stakeholders what activities they liked the most, you could ask them what activities they found the most difficult, and why. Instead of asking participants to rate ambiguous concepts like "usefulness" on a scale of more to less useful, ask them which sector or user group might find the work useful, or in which context a participatory tool might be useful.
- Draw key informants or stakeholders with a particularly close involvement in the process into one-on-one discussion about the approach during session breaks or on other occasions. A group of facilitators can usually collect a range of different reflections simply through informal conversations. One-on-one discussions often lead to constructive criticism or suggestions for improvement that would be impossible to collect from a group.
- Do not be afraid to lack confidence or show uncertainty when carrying out activities. Participants will often offer useful suggestions that lead to improvements from which the facilitator can learn. This process of continual learning by the organizers of participatory activities is a useful evaluation activity in itself.

2.1.2.4.1 Evaluation of Model Outputs

As discussed previously, one of the key goals of integrative research activities like PM is to improve communication between scientific stakeholders and other stakeholders. This is particularly important in the case of analytical models which produce numerical outputs. To increase the trust in the operation of these models and the results that they provide, it is often useful to involve a group of stakeholders in evaluating model results or outputs. These kinds of activities require careful preparation (even quite simple models often require specific training to be easily understandable to a non-specialist), but they can be very rewarding for all participants. This is easier to do with some models than others. Land-use models, for example, usually generate a map as an output, typically an artificial simulation of a date for which a real map already exists. The model is calibrated by comparing the simulation against the real map. Since the modeler usually carries out a visual assessment of similarity before employing statistical comparison techniques, this assessment phase can be carried out by stakeholders with the aid of pro-forma sheets with questions like "how much change has been correctly simulated?" (a lot, some, not much) or "how similar is the land-use pattern in the simulation to the real pattern?" (too clumped, just about right, or too dispersed) (Hewitt et al., 2014). The assessment of model calibration accuracy, particularly if carried out by a large group of stakeholders, is very useful, both to improve the model performance and to show to stakeholders how the model works. In general, a well-planned activity in which stakeholders work directly with a model is much more informative than long theoretical explanations.

If sufficient time is available and there is enough interest within the stakeholder community, it may even be possible to devolve the most important modeling decisions (variables to include, model goodness-of-fit) completely to the community, and for the model developer to take on the role simply of a technical operator or advisor. A model that is co-developed in this way is much more likely to be adopted by stakeholders as a useful tool for future planning of the territory than a "black box" which produces mysterious outputs through processes that stakeholders have never seen nor played any part in developing.

2.1.2.4.2 Participation Ladder

The participation ladder was first proposed by Arnstein (1969) as a means of evaluating the degree to which citizens are involved in a process which affects them. Geilfus (2008) uses the participation ladder to illustrate the process by which a community can take control of their own destiny, step by step, in a gradual transition from "passive spectator (beneficiary) into the driver of its own process (an agent of self-development)." Though Geilfus is primarily concerned with rural development, the participation ladder, like all the techniques described in this book, can be usefully applied in almost any context where an opportunity emerges for stakeholders to transform themselves into agents of social change. For example, the process by which a small community becomes organized to resist the imposition of a large infrastructure project, like a dam, a pipeline, or a road, can be seen in terms of steps along a participation ladder. The capacity of bottom-up movements to bring about real change is well known. In Torrelodones, Madrid, a neighborhood association first established to fight against an unwanted urban development built on its success to form a political party which was eventually elected to govern the municipality (Cueto & Muñoz, 2016; New York Times, 2011).

The participation ladder is a diagrammatic device in which the road to greater citizen empowerment is represented as a series of steps up the rungs of a ladder, the stairs of a stairway, or some other device to imply progressive ascent. If the metaphor of a journey is important to the group, a road or railway might be more appropriate, as long as it is clear that movement in one direction (up the ladder or stairs, left to right along the road) indicates progress toward the stated goal.

Each step indicates the degree of stakeholder involvement, and is accompanied by a short description of what that particular step entails. Geilfus, for example, describes step 1 (passivity) as "participation on instruction but no influence on decisions or project implementation," while step in 2 citizens participate as information providers but have no say in how the data is used. In the Doñana natural protected area we used the participation ladder to help stakeholders evaluate their degree of involvement in the development of a participatory model (Hewitt, Hernández Jiménez, et al., 2017). Here the steps progressed from, at the bottom of the ladder, simply informing key stakeholders (land planners and other protected area decision-makers) about the model upwards through stages of greater involvement of the stakeholders in model development. The top of the ladder, which was not attained, corresponded to day-to-day use of the model by the stakeholders themselves for decision-making in the protected area.

The participation ladder is a useful concept for helping social actors assess the extent to which their own involvement has grown as a participatory process has progressed, and is thus a useful proxy for understanding the level of empowerment that a process has achieved. It is also helpful more generally as an analytical tool to evaluate the degree to which real participation is being attained (as opposed to one-way information exchange) and as an idealized model of development of a participatory process. In the same way as future scenarios help to define realities we wish to attain, the participation ladder helps to specify the key elements we need to increase the autonomy of the participants in the process, to help ensure that the work has a secure and long-lasting legacy.

2.1.2.4.3 The Dartboard Technique

The dartboard technique is an easy-to-use and intuitive method to allow participants to evaluate the success of a workshop, project, or process. The technique involves drawing a circular target on a blank wallchart or other suitable medium; like in archery, the target is subdivided into consecutive rings. To give one example, the dartboard might contain just a small inner circle and a larger outer circle. Participants then evaluate the quality of the activity by marking the target using a pen, sticker, or pin, with the central circle representing "on target" (i.e., positive rating), the outer circle representing "room for improvement," and outside the target being "wide of the mark" (a negative rating—see, e.g., http://evaluationtoolbox.net.au/index.php?option=com_content&view=article&id=38&Itemid=145). To evaluate more than one element of a process simultaneously, the dartboard can be divided into quadrants or wedges (as in a game of darts), in which each wedge or quadrant represents a different activity or element of the process to be evaluated.

Alternatively, the concentric rings might not be drawn at all, allowing participants more liberty in evaluation of the activity, such that the closer the point is to the centre of the target, the higher the score. The technique has the advantage of being highly visual and easy to understand, but has the disadvantage of not being anonymous, so there is a good chance that individual scores will be influenced by the group. One way of alleviating this problem is to supply individual participants with a dartboard application for laptop, tablet, or smartphone which passes the individual scores via the internet to the facilitator, who can then collect all the scores on a master dartboard to display to the group. Such an application does not yet exist to our knowledge, but would not be difficult to develop.

As we noted above, we warn against asking participants to rate openly whether they felt that a process or activity was good or bad, since the answer is likely to depend more on the context, the social situation, or the way a particular person is feeling at the time than on the activity itself. If used carefully, the dartboard technique provides a means to allow participants to express their views without fear of being seen as too critical (or perhaps even too uncritical) and to allow various elements of the process to be compared. For example the technique might be used to allow participants to score "the most difficult activity" or "the most thought-provoking activity" (see http://www.geogra.uah.es/duspanac/3taller3res2.html, in

Spanish). It can also be used to assess if a process or method has captured the imagination of participants by asking them to score its relevance to particular sectors or target groups, with the dartboard wedges or quadrants representing these sectors or groups.

Once participants have completed the dartboard, the results can be evaluated openly by the group. It may be possible to identify clusters or patterns. The activity can be followed by a group discussion around the difference in scoring for the individual activities or elements, exploring the reason for these results. The main ideas that emerge from the debate are noted for further analysis.

2.1.2.4.4 Refutation and Analysis of Failure
2.1.2.4.4.1 Refutation

This technique has its origins in Karl Popper's *doctrine of falsification* (Popper, 2005 [1935]), a theoretical approach for differentiating between empirical statements and beliefs on the basis of whether they can be shown, in theory, to be *false*. Manufacturers use this approach when they test their products to destruction before deploying them in real-world situations, recognizing that it is not possible to say with certainty that a product will not fail under particular circumstances unless we have already made the product actually fail. Popper advocates testing scientific theories to destruction to establish their usefulness and identify their weaknesses. Here, we advocate testing participatory processes to destruction for the same purpose. Refutation can be usefully applied to almost any participatory process, activity, or model, and in practice would take the form of a group activity in which we aim to identify all realistic circumstances under which the process can be considered to have failed. Though this activity has a strong evaluative component, it is better undertaken toward the beginning of a process than at the end, since the results of the exercise may give rise to radical changes in the way we think about the process we wish to embark on.

In a refutation activity, we begin by writing the process goal. Next, by means of a group brainstorming process, we write down our starting assumptions and, next to them, the contrary position. From the contrary position, we can then brainstorm ideas that explain how this negative proposition can come to pass. For example, if we set out to change the way land planning takes place in a municipality by means of a participatory municipal plan, we assume on some level that there is at least some chance that the plan developed by citizens will actually be implemented, even if only in the sense that an official responsible for planning changes her perception after becoming aware that citizens are asking for something quite different from the *status quo*. Thus our starting assumption is "the plan will achieve some degree of implementation," and the negative proposition is therefore "the plan will not be implemented." If the result of the brainstorming exercise is "there is no conceivable method by which such a plan can be implemented," we can then modify our starting assumptions to look for other indicators of success, for example "citizens become aware that their own vision of the territory is different to that of the current administration." By this simple exercise we can transform a process from development of a participatory decision support tool that will never be used to support any decisions (like so many decision support tools) into development of a bottom-up movement for political change through the lens of land planning. While the participatory activities we undertake may be essentially the same, the chances of success are greatly increased, and as a result participants have a more rewarding experience. Instead of pouring their energies and frustrations into a policy support tool only to find the door of the town hall firmly shut in their faces, they expend the same energy in the creation of a political movement, take possession of the town hall, and implement their own participatory spatial plan themselves.[14]

The principal benefits of a refutation activity are as follows.

- It can help us change direction or emphasis with minimum cost, as outlined in the above example.
- It provides a structure for seeing what can go wrong and helping find ways to avoid this.
- It openly introduces the idea of failure to the participants, allowing the group to come to terms with the possibility.
- It is easy to become so involved in participatory processes that we begin to inhabit a bubble of self-affirmation. "Of course this will be a success," we say. "Everyone here is so nice, we all work so hard, and we agree about nearly everything." This can lead to great disappointment when we do not see the radical changes we expect take place. The kind of detailed information provided by a refutation exercise will often show us why we are unlikely to succeed in the way we expect (e.g., because our starting assumptions were wrong)
- It allows us to separate tried-and-tested approaches from nice-sounding theories, and focus on what can really be achieved. For example, we may be more successful at "developing a community wind turbine" than "achieving a sustainable energy transition." This is not to say that idealistic or optimistic proposals should never be considered (quite the contrary: to achieve real change, we must "dare to dream"—see Section 2.1.2.3.3 on scenario development), but only that we need to be realistic about what we can do with the time and resources available. It can be very damaging to set overambitious goals,[15] since participants may become disillusioned and withdraw their engagement.

2.1.2.4.4.2 Analysis of Failure

Analysis of failure is not a specific technique, but rather an evaluation process that could include a range of other techniques already discussed in this volume. Analysis

---

[14]Skeptical readers are reminded that this actually happened (see Cueto & Muñoz, 2016; New York Times, 2011), and, we firmly believe, will continue to happen as long as land planners and policy-makers continue to pay homage to the (conclusively falsified) paradigm of growth without limits at the expense of the territory (see Chapter 1).

[15]Ambitious goals, can, however, be broken down into manageable chunks and assigned to WGs (see Section 2.1.2.3.5). If some of these groups fail to achieve their objectives, it will not necessarily jeopardize the whole enterprise.

of failure of a specific case or historical example might itself be a useful starting point for a participatory process cycle involving a range of techniques, e.g., interviews, discussion groups, analysis of stakeholder discourse around reasons for failure, a timeline showing development of the process to identify key historical decisions that are thought to have led to failure, and problem/objective trees to identify pathways explaining failure. This kind of activity is especially useful in situations where there is extensive disagreement among the stakeholder community about what to do. It may be necessary to manage participatory activities carefully to avoid conflict, but the results might be revealing—do different actors actually blame each other? Or is there a problem of a communication? Academic analyses of failures of participatory processes are not easy to find but make worthwhile reading (Daley & James, 1992; Nielsen, Fredslund, Christensen, & Albertsen, 2006).

2.1.2.4.5 Analysis of Viability of Action Plans

Following the development of action plans in the action phase (see Section 2.1.2.3.6), it is desirable to carry out a thorough analysis of their viability. Ideally, this would be done before putting the actions agreed into practice. Alternatively, viability analysis could be carried out a short while after beginning implementation as a means of critically evaluating progress. No specific method can be recommended, but the viability analysis should certainly be carried out by as many of the participants of the WGs tasked with implementing specific action plans as possible, and with a minimum of intervention by facilitators or other external stakeholders. One possible approach would be for each WG to analyze the viability of another WG's action plan, and then share the results in an open session. The main aims of the viability analysis are as follows.

- Identify the specific interventions (tasks) for implementation of each action plan.
- Establish a schedule for the implementation of the defined interventions. This need not be strict, but should at least try to set attainable milestones.
- Determine the human resources and materials which are necessary for carrying out the actions.
- Assess the possibilities of continuation of the work initiated with the key stakeholders in the action phase, e.g., MGs, WGs, and facilitators.
- Strengthen the transfer of collective leadership by clarifying that responsibility for success or failure of the defined actions lies with the stakeholder community.

As one of the last stages in which external facilitators are involved, it might be useful to dedicate a whole event or workshop to viability analysis and try to bring together all the social agents that have been involved in the process so far, as a means to encourage cooperation and joint action.

Experiences

3

In Chapter 3 we present a collection of case studies showing the way the techniques and approaches described in detail in the preceding chapter have been applied in real situations. Since the case studies are very diverse, they have been grouped into three key themes.

Theme 3.1: Getting to know the territory: in dialogue with the past, understanding the present, thinking about the future

Theme 3.2: Between city and country: building more resilient rural–urban relations

Theme 3.3: Conflicts, citizens and society: participatory modeling for a resilient future

3.1 THEME 3.1: GETTING TO KNOW THE TERRITORY: IN DIALOGUE WITH THE PAST, UNDERSTANDING THE PRESENT, THINKING ABOUT THE FUTURE

INTRODUCTION

The territory, with its landscapes, fields, villages, and farms, is not just a product of the human populations that dwell there and till its soils or drive animals through its pastures. It is shaped by the complex interplay of multiple individuals and organizations from far and wide. These include global trade consortia, large-scale food suppliers, supranational bodies, national governments, province or state regulators, farmers' trade unions, local planners, conservation organizations, and local businesses. The number of these *social agents* or *stakeholders* who seek to influence the territory has increased significantly compared to earlier times. At the same time, the connections between the different stakeholders and the territory, and between the stakeholders themselves, have become more complex and indirect.

The approach we put forward in this book seeks to unpick this complexity and navigate this diverse community of actors such that the knowledge and traditions of local stakeholders, those who live in and best know their territory, provide the foundation for actions that rise higher up the chain. This is the opposite of the way that most land is managed today—where the most important decisions about water, air, soil, and agriculture are made remotely by actors with no connection to the places in which their effects are most strongly felt, and are taken without any

Developments in Environmental Modelling, Volume 30, ISSN 0167-8892. https://doi.org/10.1016/B978-0-444-63982-0.00003-3

consideration of the values, interests, and identities of local landscapes and communities.

But the locally-based governance approach that we propose as an alternative to this disorienting spider's web of remote decisions and directives is not possible without a profound knowledge of both the territory and its inhabitants. We need to know how the territory has evolved over time, and how natural processes and human activities have come together to shape it. We seek to understand the imprint left in a particular place by its inhabitants throughout the centuries: in the physical form this may take, such as fields, quarries, buildings, and paths, in soils, plants, and animals, and in less tangible aspects, like the rich store of local knowledge about a place and its rhythms, traditions, customs, and practices that may still be preserved in the minds of its present-day inhabitants. Often this knowledge may lie deeply hidden, accessible only to archeologists, historians, and of course local people, who may need some encouragement to reveal their store of information. Older people, in particular, may feel their traditional knowledge has no value, or that no one is interested in hearing it. Thus the first step along the road to building a better future, in line with the principles of sustainability and social justice, is to understand better and value what we already have.

The first part of this chapter addresses this key theme of the collective appreciation and revitalization of our cultural and natural heritage and common spaces by local communities. In the case studies described here a wide range of approaches and methods are employed to link the past with the present and help local stakeholders develop proposals for the future. The proposals that emerge in this way, based on a deep knowledge of past and present land-use approaches and land working practices particular to each place of origin, are strongly culturally embedded and stand a greater chance of implementation than proposals that come from outside. The following specific cases and initiatives are described in this section.

Case 3.1.1: Baja Alcarria droveways strategy: in dialogue with the past. Funded by Departamento de Vías Pecuarias, DG Agricultura y Ganadería, Regional Government of Madrid (2009—2010).

Case 3.1.2: Participatory strategy for the revitalization of traditional movements of livestock in natural areas of Madrid and Cantabria. Funded by Fundación Biodiversidad, under the call for grants on biodiversity, climate change, and sustainable development activities (2011—2012).

Case 3.1.3: Recognizing the heritage value of rural public works. Difficulties, opportunities, and challenges (VAPROP). Funded by Red Rural Nacional and coordinated by Fundación Miguel Aguiló. Observatorio para una Cultura del Territorio (OCT) collaborated in the participatory process and the landscape characterization alongside the Ecology and Landscape Research Group of the Universidad Politécnica de Madrid (2009—2014).

Case 3.1.4: Participatory strategy for the revitalization of public space around the Jarama River (Madrid). Collaboration with the Consumo-Gusto (Ciempozuelos) and The Association for the Recovery of Autochthonous Woodland (ARBA) (2012—present).

FIGURE 3.1

Cattle drinking trough from the droveway between Colmenar Viejo and El Boalo (municipalities of Madrid).

Case 3.1.5: Common grazing and transhumant livestock management in Hermandad de Campoo de Suso municipality (Cantabria). Funded by Asociación País Románico (2011).

Case 3.1.6: Project assessment: food sovereignty consolidation in Coumbacara, Senegal. Agencia Andaluza de Cooperación Internacional al Desarrollo. Project coordination: Paz y Dignidad association. Project implementation: local partner 7° and National Association for Adult Literacy and Training (2015).

Case 3.1.7: Cattle droveways, pathways to history. Landscape and heritage along the droveways network (Fig. 3.1). In collaboration with Área Arqueología cooperative. In progress (2017).

Case 3.1.8: Public participatory geographic information systems: adolescent perception of landscape elements, values, and changes in Colmenar Viejo (Madrid). On behalf of the University of Hohenheim, Germany, under the HERCULES Project, funded by the Seventh Framework Programme of the European Union (2016–2017).

GENERAL APPROACH AND STRUCTURE

In this section we present the principal tools and approaches that appear in the case studies described in Theme 3.1. Fig. 3.2 highlights the tools that are most relevant and most frequently applied in this theme. These techniques are described in detail in Chapter 2. In Section 3.1.3 we present the case study projects listed above to show how these techniques have been applied in practice in a diverse range of situations.

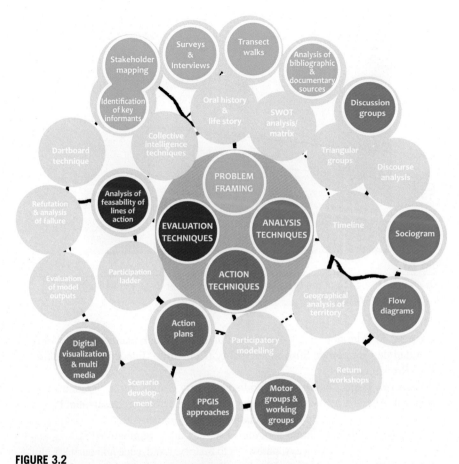

FIGURE 3.2

General methodological framework for Theme 3.1. The techniques which we have most frequently used in this theme are highlighted.

The techniques used in this theme fall into two broad groups: those that are directed toward recovering the socioecological and cultural memory of the territory (essential for building a common future), and those focused on participatory characterization and design of new proposals to turn this collectively imagined future into reality. Thus in the former group we find appraisal techniques like *stakeholder mapping and identification of key informants* (Chapter 2, Section 2.1.2.1.1) and *transect walks* (Chapter 2, Section 2.1.2.1.4), and analysis techniques like *sociograms* (Chapter 2, Section 2.1.2.2.2) and *discourse analysis* (Chapter 2, Section 2.1.2.2.6). In the latter group we find analysis techniques like *flow diagrams* (Chapter 2, Section 2.1.2.2.5) and *return workshops* (Chapter 2, Section 2.1.2.2.8), and action techniques like *motor groups* (Chapter 2, Section 2.1.2.3.5) and *action plans* (Chapter 2, Section 2.1.2.3.6). In the course of a project we see how these techniques are applied in succession, as for example in *Case 3.1.2: Participatory strategy for the revitalization of traditional movements of livestock in natural areas of Madrid and Cantabria*, where *sociograms* are employed to understand the stake-holder community and their relationships, *transect walks* are used to explore the territory and interview the users of the droveways in an informal situation, *flow diagrams* are employed to analyze problems and develop actions to solve them, and finally these actions are subject to a comprehensive *viability assessment* by the stakeholders themselves. Fig. 3.2 shows all the techniques used in this theme; Fig. 3.3 shows how each technique relates to a specific project.

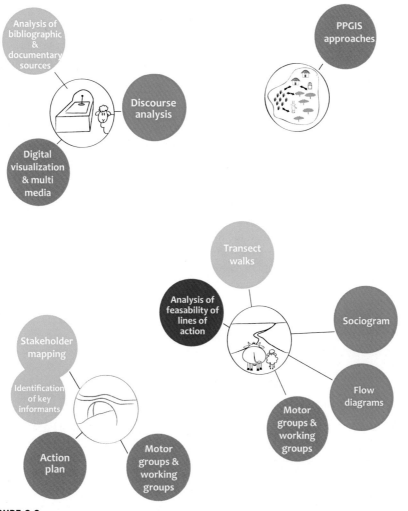

FIGURE 3.3

Diagram showing the relationship of the techniques used, following the color coding used in Figure 2.2 (Chapter 2), to the projects described in this theme. Each project is given a symbol, which appears in the top right-hand corner of each factsheet to help the reader identify the project.

DESCRIPTION OF PROJECTS AND TECHNIQUES USED

This section presents a synthetic review of each project included in *Theme 3.1: Getting to know the territory*, with a brief description of the techniques employed in each case. This review is presented as factsheets containing basic information about each project, its geographical location, context, funding body, and main objectives, and a synthetic description of the work carried out. The factsheets are intended to provide a rapid appreciation of the characteristics of each project to enable the reader to understand how each particular case determined the choice of techniques applied and the most important benefits derived from their use.

It is important to stress that the techniques described in the factsheet for each project are not necessarily the only techniques employed in that particular case study. For example, analysis of documentary sources has been used in almost every case; this is, however, described only once, in *Case 3.1.1, The Baja Alcarria droveways strategy*. Sometimes techniques described in Chapter 2 appear as part of another technique or activity—for example, the establishment of working groups (WGs) was a key part of the workshop on "Analysis of viability of action lines," part of *Case 3.1.2: Participatory strategy for the revitalization of traditional movements of livestock in natural spaces of Madrid and Cantabria*, but these are not discussed separately. To do so would be repetitive, break up the flow of the narrative, and, worse still, encourage the idea that the participatory processes described are no more than the sum of their parts, a collection of techniques thrown together for a particular aim. Rather, the cases described each developed their own objectives and distinctive visions as the work progressed, and the methods used tended to arise organically as a response to these objectives and visions. We make no apology for this "mix-and-match" approach: the aim of these case studies is to show the application of participatory research approaches, as we understand them, in the real world.

3.1.1 CASE 3.1.1: BAJA ALCARRIA DROVEWAYS STRATEGY: IN DIALOGUE WITH THE PAST[1] (2009–2010)

3.1.1.1 Project Synthesis

This project was carried out in the Baja Alcarria area in the Madrid region, mainly in the municipalities of Pezuela de las Torres, Olmeda de las Fuentes, Ambite, Orusco de Tajuña, Brea de Tajo, Valdaracete, Estremera, Villarejo de Salvanés, and Fuentidueña de Tajo. These are crossed by a great number of livestock paths of different levels of importance, from simple pathways known as *cordeles* to royal droveways known as *cañadas*, which are crucial communication routes of great historic importance.

The objective of the project was to valorize the rural areas linked with the droveways by developing an understanding of the past as a way to know the present and

[1]Funded by Departamento de Vías Pecuarias, DG Agricultura y Ganadería, Regional government of Madrid.

imagine the future possibilities of these spaces. The first stage of the project, named "Dialogue with the past," was mainly focused on historical analysis, in particular understanding the evolution of drovers' roads, transhumance, and livestock farming in the area, as well as their role as driving forces of local development in the region. The main lines of action were identification and selection of key stakeholders in the area; the search for and consultation of literature and relevant documents about droveways; and collection of oral histories from local inhabitants.

These actions were undertaken concurrently with a dissemination program named "Discover your droveways" carried out by the regional government, which included separate dissemination and valorization activities. Subsequently we also participated in a European research project, CANEPAL (www.prismanet.gr/canepal/), which aimed to bring together all the European data sources about pastoral life and transhumance, transport and routes, herd management, architecture related to sheep-farming activities (Fig. 3.4), cultural heritage, and linked social activities. Although these projects are now complete, we maintain links with social organizations that work in defense of public paths in the Madrid region (Ecologistas en Acción and the Sociedad Caminera del Real de Manzanares), with the aim of ensuring continuity of the work carried out.

3.1.1.2 Description of Techniques

3.1.1.2.1 Analysis of Bibliographic and Documentary Sources (Chapter 2, Section 2.1.2.1.6)

Literature review and analysis of secondary sources were used in this project to understand about livestock and its management in this area in the past. Information drawn from secondary sources was analyzed in two stages: first, the information was read and assimilated; and second, a "situation report" was prepared based on this information. The information collected was organized in three sections: general

FIGURE 3.4

Old stone-built shepherd's shelter, located in the Baja Alcarria.

transhumance; Madrid—Alcarria; and cultural heritage. Information in each section was classified as essential, interesting, and peculiar, so a reader of the report could prioritize according to her or his interest.

The following historical sources were consulted.

- Archives of Asociación de Ganaderos del Reino (National Historical Archive).
- Archive of Precedents of the Madrid region (Comunidad de Madrid).
- Archive of Marqués de la Ensenada.
- Forest Goods and Heritage Service (S.G. Medio Rural y Agua).
- Library of Asociación Trashumancia y Naturaleza.

3.1.1.2.2 Discourse Analysis (Chapter 2, Section 2.1.2.2.6)

Discourse analysis of information collected from primary sources involved transcription of material and content analysis. Two different types of transcripts were made.

- Literal transcription: used for all of the stakeholder interviews and some of the stakeholder meetings.
- Summary of contents: used for some of the stakeholder meetings and for informal discussions and conversations.

The objectives of the analysis were as follows.

- Defining the social stakeholders' views, aims, and strategies.
- Building/improving the social map.
- Identifying historical milestones.
- Improving the qualitative and quantitative data collected from other sources.
- Triangulating the qualitative and quantitative data collected from other sources.

3.1.1.2.3 Digital Visualization and Multimedia (Chapter 2, Section 2.1.2.3.4)

The visualization materials were quite simple due to the limited resources available for the project, and comprised a series of attractively designed posters for display at the final workshop (Fig. 3.5).

3.1.2 CASE 3.1.2: PARTICIPATORY STRATEGY FOR THE REVITALIZATION OF TRADITIONAL MOVEMENTS OF LIVESTOCK IN NATURAL AREAS OF MADRID AND CANTABRIA[2] (2011—2012).

3.1.2.1 Project Synthesis

This project was developed in the regions of Madrid and Cantabria, in the Cuenca Alta del Manzanares Regional Park and the Oyambre and Saja-Besaya Natural Parks (Fig. 3.6), respectively.

[2]Funded by Fundación Biodiversidad, under the call for grants on biodiversity, climate change, and sustainable development activities.

Theme

FIGURE 3.5

Dissemination posters picturing the culture and history of transhumant pastoralism, droveway routes, and landscapes.

The main objectives of the project were to conduct a participatory characterization of the state of the livestock routes[3] and develop a participatory strategy to promote their use and enhance public awareness. This involved searching for consensus between local users and the land management authorities to develop a use and management approach that was compatible with sustainable use of these routes by livestock. This approach integrates the maintenance of knowledge, innovation, and local practices, protection of ecosystems and biodiversity, and the economic viability of the livestock farming sector in the research areas. In this way, revalorization is linked

---

[3]In this project the term "livestock routes" (*rutas ganaderas*) was preferred over "droveways" (*vías pecuarias*). This is because the term droveways implies protection under Spanish law, which not all such routes necessarily enjoy.

FIGURE 3.6

Livestock grazing in Cantabria.

with revitalization of the traditional use of the livestock routes as a potential source of economic activity. During the project several participatory processes were conducted in municipalities in the research areas, which resulted in the following outcomes.

1. A participatory strategy for the revitalization of traditional livestock routes in preserved natural areas.
2. A new line of intervention to promote the use of livestock in fire prevention and forest management, as well as for rural economic revitalization and preservation of common and protected areas.

More information about the results of this project can be found at http://observatorioculturayterritorio.org/wordpress/?page_id=175.

The work was carried out in four principal stages.

1. Identification of the stakeholder community (sociograms).
2. Understanding the territory together with key stakeholders (transect walks with livestock keepers).
3. Development of lines of action through workshops for definition of action lines in each region (Madrid and Cantabria) using flow diagrams and a problem−action matrix.
4. Follow-up workshops to assess the viability of lines of action previously developed (flow diagrams and a problem tree).

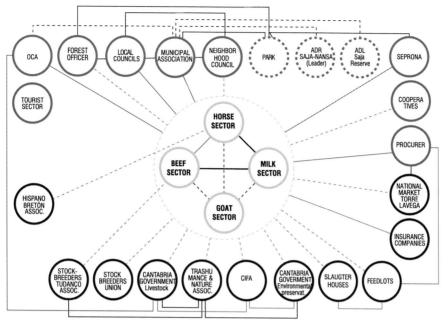

FIGURE 3.7

Preliminary sociogram for the study area, Cantabria.

3.1.2.2 Description of Techniques
3.1.2.2.1 Sociogram (Chapter 2, Section 2.1.2.2.2)

Three different types of sociogram were prepared in this project. A preliminary sociogram was made on the basis of the initial review of secondary sources. This depicted the key stakeholders with knowledge about or involvement in this territory, and the relationships between them (Fig. 3.7). The lines relate to different types of connection between the stakeholders, as shown below.

Stakeholders	Connections
Green circle/box: not organized	**Continuous black line**: existing (habitual) relationship
Blue circle/box: organizations	**Dashed black line**: occasional relationship
Dashed blue circle/box: organizations with less involvement in the issue	**Continuous red line**: conflict relationship
Dashed green center circle: set of non-organized participants	
Red circle/box: institutions	

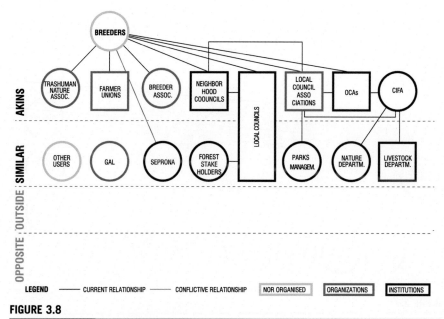

FIGURE 3.8

Positioning sociogram for the study area, Cantabria.

On the basis of the information obtained from the individual interviews, we developed more detailed sociograms, the so-called individual sociograms, in which each researcher expressed his/her own personal understanding of the relationships between actors. Finally, these sociograms were modified according to information collected in the workshops to include the current links between the stakeholders at the time the workshops were carried out and their degree of affinity to the project's aim. This was important to identify which actors to work with in future stages of the project.

This final version is known as a positioning sociogram (Fig. 3.8). The same legend (see above) was used to depict the relationships. The sociograms were instrumental in understanding which actors were relevant to the process at hand and the relationships between them (some clear, some hidden, some emerging). For successful development of actions that stand a realistic chance of implementing genuine change, correct identification of the actors involved and their relationships is essential.

3.1.2.2.2 Transect Walks (Chapter 2, Section 2.1.2.1.4)

Fig. 3.9 shows the route taken by participants on a transect walk which followed a droveway. Information extracted from the discourse of the informants is shown on the map, including the main features associated with the droveway, like drinking

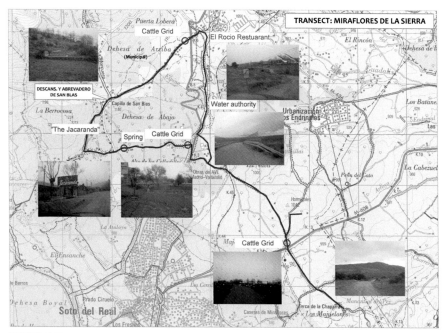

FIGURE 3.9

Miraflores de la Sierra—Soto del Real transect map.

troughs and resting points, their use, associated problems, and potential for future development. The transect walk (Fig. 3.10) was found to be particularly useful in this case for identifying existing conflicts experienced by the farmers in the use of droveways (mainly encroachment). A large amount of information also emerged about land-use changes in the area, the farmers' discourse about extensive livestock farming, and the current use of the droveways among other aspects, as the following excerpt illustrates.

Miraflores de la Sierra—Soto del Real transect description excerpt

The beginning of the transect is located at the junction between the M-611 road and the Calleja de las Suertes drovers' path. A cattle crossing point has recently been created. Previously it was not possible for the cattle to follow this route, and farmers had to cross the road through a problematic spot at which there is a change of slope.

The itinerary continues past the previous livestock crossing point and along the path toward the valley of the Saelices brook. Before reaching the railroad this path joins the Calleja de las Suertes drovers' path, which is wide and well-maintained because the route was used by service vehicles for the high-speed train. Although it is an easy route for the herders, cars can drive down here very fast, so problems may arise when they meet the cattle.

FIGURE 3.10

Photograph of one part of the transect.

The route follows the Calleja de las Suertes drovers' path toward the valley, taking a wide bridge over the railway line. This bridge is protected with crash barriers that can produce cuts to the cow's legs; a different design would be more appropriate. Also, the fences could be improved here to stop the cattle getting off the bridge.

The farmer explains that the Calleja de las Suertes drovers' path links the valley and the mountain pastures, so local farmers use it to move the cattle from the winter to the summer pasture areas. However, he doesn't take cows to the mountains himself, but he uses this route to take them between his properties. The reason he doesn't take them to the mountain pastures is because he's been practicing organic livestock farming for the last 4 years (not certified), and needs to control what the cows are fed. He does not sell the meat as organic, because of the costs associated with obtaining the official certification. For instance, organic cattle need be taken to slaughterhouses in Avila or Segovia, which are prepared to handle organic meat. Currently he sells his meat as Beef from Guadarrama (a protected geographical denomination) in Hiber (a supermarket chain), and is planning to go for full organic certification in the future.

3.1.2.2.3 Flow Diagrams (Chapter 2, Section 2.1.2.2.5)

Two flow-diagram approaches were used: a problem—action matrix developed with all stakeholders during the action lines definition workshops to identify problems, prioritize the critical points that prevent action from being taken, and develop action lines to address them; and a problem tree prepared by the research team to assess the viability of the lines of action proposed prior to the workshops on this topic.

Problem–Action Matrix

This technique was applied as part of the action lines definition workshops held in both Madrid and Cantabria.

Users of droveways and public paths attended the Madrid workshop, and the livestock keepers of Cuenca Alta del Manzanares Regional Park were well represented among the participants. In Cantabria participants included livestock keepers, butchers, and representatives of farmers' organizations.

The practical application of the technique in the project can be summarized as follows.

First, participants received a post-it note instructing them to write down the factors they considered to be related to the issue addressed. The main ideas were grouped by thematic areas and transferred to a flipchart. Stakeholders then identified the critical points (see Chapter 2, Section 2.1.2.2.5.1).

Subsequently, according to the problem areas defined in the problem matrix and following discussion about the capacity to resolve these problems, five action lines were established (Fig. 3.11).

1. Addressing problems related to users and lack of public awareness about the use of droveways, and about the environmental, social, and economic benefits of livestock and livestock movements for the territory.
2. Addressing livestock movement problems related to use and management of natural resources associated with extensive livestock farming.
3. Addressing livestock movement problems related to lack of organization, support, and coordination in the extensive livestock farming sector, as well as its coordination with other users and competent authorities.

FIGURE 3.11

Stakeholders study the action lines developed using the problem–action matrix.

4. Addressing livestock movement problems related to the low annual profitability of the livestock farming sector.

5. Addressing livestock movement problems related to the use and management of the droveways.

Problem Tree

The problem tree was prepared by the research team based on the results of the first workshop. We grouped the main problems of the sector in areas, from the most general problems to those specific for each line of action. All the problems prioritized by stakeholders in the first workshop were linked with cause-and-effect arrows. The problem tree was intended to provide an initial overall picture of the current situation for participants to amend and complete during later workshops. The final problem tree is shown in Fig. 3.12.

3.1.2.2.4 Analysis of Viability of Action Lines (Chapter 2, Section 2.1.2.4.5), and Working Groups (Chapter 2, Section 2.1.2.3.5.2)

The action lines agreed in the previous workshop were subject to participatory analysis of viability in dedicated workshops in Madrid and Cantabria. These workshops had the dual objective of sharing with participants the principal results of the participatory process obtained so far (see Chapter 2, Section 2.1.2.2.8).

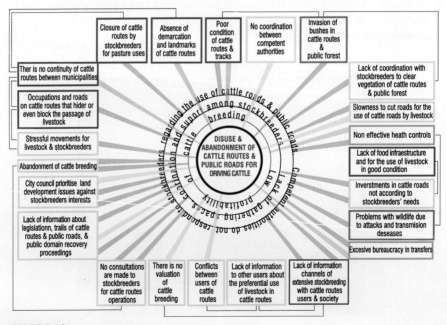

FIGURE 3.12

Problem tree for revitalization of traditional livestock movements in Madrid and Cantabria.

Before the workshops, participants were asked which action line they wanted to participate in, whether they already had any preliminary proposals or specific issues to discuss in the workshop, and what their potential involvement might be in any such proposal. In addition, the following information was sent to each participant.

1. A document of general issues, including the results of the participatory process carried out up to that time.
2. The problem tree prepared by researchers for the participants to amend and complete.
3. A document of potentialities and problems, setting out the findings of the participatory research carried out up to that point in time.

Four WGs were formed to facilitate discussion and proposal development (two in Cantabria, two in Madrid). Once the WGs were organized, rapporteurs were assigned to moderate the discussions and present the reflections and conclusions emerging from each WG (Fig. 3.13). Each WG was assigned three action lines to work on (one action line was assigned to both groups in both areas), and tasked with developing proposals to ensure long-term viability. Once the group work was finished, the proposals developed by the WGs were presented in a plenary session for discussion and reaching an agreement on further work.

When the proposal of WG A was presented in Madrid, most participants on WG B were interested in developing it, so a new WG was formed and the participants in this group agreed to meet again beyond the end date of the project to develop the proposal. The actions proposed by the two groups in the Madrid action lines viability workshop are shown in Table 3.1.

FIGURE 3.13

Working session on prioritized actions.

Table 3.1 Actions Proposed by Participant Groups

Working Group	A	B
Specific proposal	Develop mechanisms to support farmers of the upper basin of Manzanares regional park to prepare a report on infrastructural needs in the cattle routes of the zone and deliver it to the competent administrative authority	Promotion of the quality of the livestock product
Proposal details	A second objective was to continue creating spaces where the different users could continue to meet and pursue specific work to support the livestock sector that was considered key for the protection of livestock roads in Madrid To this end, a date was established for a next meeting, and the actors currently involved in this initiative and those not present but necessary to involve were identified One participant agreed to act as a facilitator to convene this next meeting	A second objective was to promote local products, specifically educating consumers about the benefits of local food and livestock production, and to strengthen short-distance commercialization networks in the area to market products of the regional park directly to consumers One member of this group agreed to act as a facilitator for this line of action The group also identified the key actors already involved in this initiative, as well as those not present but necessary to get involved in it

3.1.3 CASE 3.1.3: RECOGNIZING THE HERITAGE VALUE OF RURAL PUBLIC WORKS. DIFFICULTIES, OPPORTUNITIES, AND CHALLENGES (VAPROP)[4] (2009–2014)

3.1.3.1 Project Synthesis

The project addressed valorization of the rural heritage of public works (roads, bridges, public buildings, etc.) in three rural counties chosen as pilot regions: Extremadura, Cantabria (Fig. 3.14), and La Rioja. Interpretation of these monuments in their landscape context from the point of view of cultural heritage is an

[4]Funded by Red Rural Nacional and coordinated by Fundación Miguel Aguiló. OCT acted as a collaborator in the participatory process and the landscape characterization alongside the Ecology and Landscape Research Group of the Universidad Politécnica de Madrid.

FIGURE 3.14

Royal droveway in the Besaya corridor (Cantabria).

important first step in developing the identity of public works beyond their physical appearance or daily use.

The study applied an innovative methodological process aimed at securing the involvement and support of local communities at all stages of the project. A comprehensive analysis of the territory was carried out using integrative research approaches which combined quantitative analysis of socioeconomic data with qualitative methods, like interviews and workshops with local stakeholders, to gain a better understanding of the public works in their socioenvironmental context.

Empowerment, co-learning, and knowledge transfer were among the key aims of the project. In particular we were interested in promoting the value of public works for local communities as a rural heritage resource with potential to stimulate local development. Dissemination activities were thus aimed at raising awareness of the value of local heritage as a cultural resource linked to collective identity.

The problem of the wide dispersion of public works in these rural areas was addressed in an innovative way through new technologies, especially internet-enabled mobile devices. Heritage routes, including information and precise directions for following appropriate paths, were prepared to enable visitors to access and interpret the sites using smartphones or other devices.

3.1.3.2 Description of Techniques
3.1.3.2.1 Stakeholder Mapping and Identification of Key Informants (Chapter 2, Section 2.1.2.1.1)

The project was carried out in three rural areas in different regions, Extremadura (Valencia de Alcántara), Cantabria (Campoo Los Valles), and La Rioja (Sierra Cebollera), while the research team was based in Madrid. Stakeholder mapping (Fig. 3.15) was therefore primarily focused on identifying key informants able to

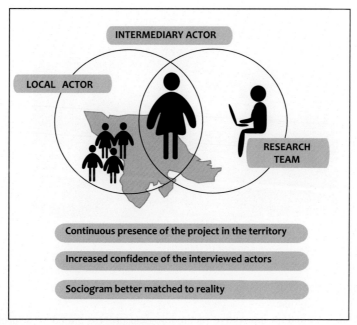

INTERMEDIARY ACTOR

LOCAL ACTOR

RESEARCH TEAM

Continuous presence of the project in the territory

Increased confidence of the interviewed actors

Sociogram better matched to reality

FIGURE 3.15

Stakeholder map for the VAPROP project.

act as intermediaries between the research team and local stakeholders, ensuring smooth communications and helping to maintain networks of trust.

3.1.3.2.2 Motor Groups (Chapter 2, Section 2.1.2.3.5.1)

The multidisciplinary nature of the VAPROP project, which brought together perspectives as diverse as local development, civil engineering, tourism, and environment, led us to try to build stakeholder groups that were able to integrate these contrasting visions—an approach which is still quite unusual in research. These groups formed the basis for the motor groups (MGs) which acted as facilitators and drivers of actions in the territory (Fig. 3.16).

The Comarca de Campoo Los Valles (Cantabria) MG was created after a three-stage process comprising stakeholder identification, stakeholder interviews and transect walks, and a strengths, weaknesses, opportunities, and threats (SWOT) analysis. The aim of this MG was to identify sites of interest for rural heritage valorization and design a common valorization strategy involving different sectors of the population.

The first meeting of the MG was held in Matamorosa in Cantabria: participants included intermediating stakeholders (key informants), local development facilitators, representatives of Fundación Tecnología y Territorio (a not-for-profit

FIGURE 3.16

Motor group meeting.

organization that works for regional development under sustainability principles, using new technologies as a tool), the Cantabria rural tourism association, Cantabria University, and a territorial development association based in Campoo Los Valles. After a series of meetings, various sites were defined: Ebro River Reservoir, the Híjar River, source of the Ebro River, the Besaya corridor, and the Camesa and La Robla railway. Subsequently a program of disseminating cultural and natural heritage values of these areas was carried out, promoted by the MG and the research team.

3.1.3.2.3 Action Plans (Chapter 2, Section 3.6)

The research team collaborated in different ways in the development of valorization actions regarding public works, depending on the specific case. One of the most successful actions, developed for the Campoo los Valles area, involved the participatory production of interpretational guides for the sites identified in the MG meetings, placing them in their landscape context. Fig. 3.17 shows one of these guides for the Híjar Valley, in which its cultural heritage sites, landscape, and communities are identified and described for visitors on the basis of information supplied by local stakeholders. The main action lines agreed by stakeholders were also given in the guide (Fig. 3.18). These include promoting future projects related to countryside stewardship and environmental education, and improving coordination of the existing tourist literature.

Unidad

VALLE DEL HÍJAR

Valle del Híjar

La estación de esquí de Alto Campoo confiere actualmente al Valle un carácter muy marcado por el sector turístico. Atrás quedaron las actividades más vinculadas al sector rural, la ganadería y la agricultura, así como la actividad tradicional de la molienda. Aún así, todavía es posible apreciar el peso del sector ganadero en la zona con el aprovechamiento de los pastos del Valle, que traen recuerdos de un pasado no tan lejano.

DATOS BÁSICOS

Superfice: 21.748 ha

Altitudes: 1300m

Mapa IGN 1:50.000: 82,3,107, 108

Núcleos: Reinosa, Néstares, Salces, Villacantid, fontibre, Riaño, Entrambasaguas, Estación Alto-Campoo. Todos los pueblos están fuertmente influenciados por Reinosa.

Vías de comunicación:
Carretera Autonómica CA183, carreteras comarcales y paseo peatonal de Násares a Entrambasaguas.

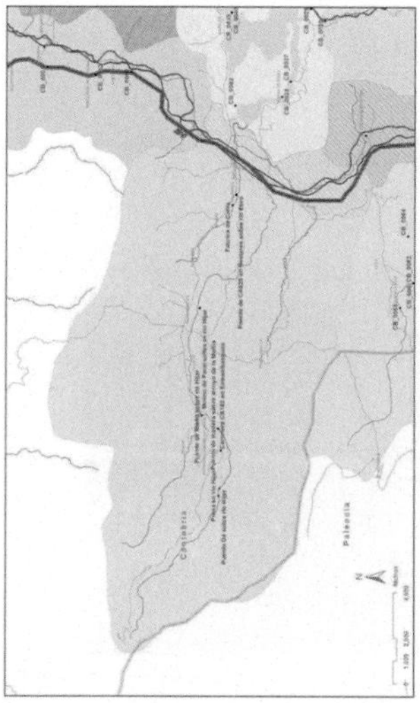

FIGURE 3.17

Excerpt from the participatory guide for the Híjar Valley: description of the landscape unit.

EL PICO TRES MARES, EL NACIMIENTO DEL EBRO Y EL TRASCURRIR DEL HÍJAR

Evolución histórica:

En la actualidad todos los pueblos de esta unidad están fuertemente influenciados por Reinosa (cabeza de comarca, cuyo peso de atracción urbana se extiende por todo el valle) acogiendo segundas residencias de población campurriana, hoy en día residente en las grandes ciudades.
El Valle del Híjar ha sido origen de multitud de aprovechamientos tradicionales, tanto el agua del riego para los regantes del valle, como para el abastecimiento de la comarca. La extracción de áridos es tradicional para la construcción, pudiendo encontrar numerosos muros de bolos en las viviendas a lo largo del Valle. También se extrajo madera para la construcción y calefacción. Una unidad fuertemente marcada por la imponente silueta del Pico Tres Mares en el Alto de Campoo, el conjunto arquitectónico en el entorno al nacimiento del río Ebro y los diferentes puentes, y numerosas construcciones hidraúlicas en el trascurrir de las aguas del Híjar.

Valle del Híjar

Obras Públicas destacadas y principales elementos culturales:

Recorriendo el valle, todavía se puede apreciar la existente trama de prados delimitados por muros de piedra seca de gran interés por su disposición en el terreno o por su tradicional uso. Igualmente, encontramos fábricas y molinos que nos recuerdan la importancia de la actividad molinera en el pasado. Aunque muchos de los molinos han sido transformados en viviendas, aún es fácil reconocer en su estructura la función desempeñada en el pasado. Hay que resaltar también los sistemas de canales de regadío. Nuevas construcciones, que en muchas ocasiones han respetado la estructura vernácula, ofrecen servicios para el turista (posadas y casas rurales), y a la par aparecen numerosas áreas recreativas en el margen de la carretera nacional, cuyo trazado es prácticamente paralelo al río Híjar y al río Ebro desde su nacimiento en Fontibre. A partir de las revueltas de Entrambasaguas, la carretera se adentra en el ambiente de alta montaña, en el que los bosques y prados dan paso a pastizales alpinos y las condiciones climáticas más duras. A lo largo de todo el camino de la Carretera Autónomica CA-183, nos acompaña el paseo peatonal de reciente construcción, que une los pueblos del valle con el núcleo con mayor peso en la zona, Reinosa.

RECORRIDO

El recorrido propuesto para conocer el Valle del Híjar tiene como punto de partida Reinosa con el Puente de Carlos III localizado muy cercano al lavadero del Río Ebro, tomamos la carretera hacia Néstares dejando a nuestro lado el paseo peatonal por el qUe caminan diariamente grupos de campurrianos. Pasamos por la cercana población de Néstares, más adelante encontramos la Fábrica de Ciella, la cuál podemos avistar desde la carretera. Más adelante hay que adentrarse en unos caminos rurales para poder llegar hasta el Molino de Paracuelles, en desuso pero en buen estado. El puente de Riaño y puente Dé sobre el Híjar destacan frente a otros puentes de menos envergadura en su construcción, pero con el mismo peso en la historia del lugar, que encontramos en el camino como el puente de madera sobre las losas erosionadas por las grandes crecidas del Híjar. Sin duda merece la pena continuar tras Entrambasaguas para disfrutar de la panorámica que nos aporta la carretera en su ascenso al Alto de Campoo.

PROPUESTAS DE ACCIÓN:
- Proyectos de Custodia del Patrimonio Cultural de Molinos, etc.
- Proyectos de Educación Ambiental
- Coordinación de las publicaciones existenes de los recursos turísticos.

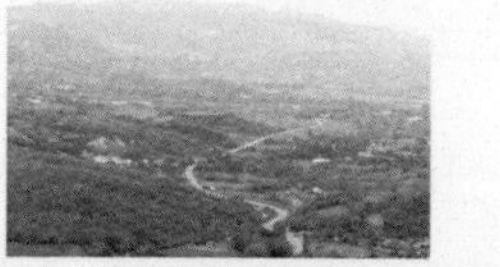

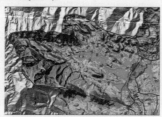

FIGURE 3.18

Excerpt from the itinerary and action proposals in the area.

3.1.4 **CASE 3.1.4: PARTICIPATORY STRATEGY FOR THE REVITALIZATION OF PUBLIC SPACE AROUND THE JARAMA RIVER (MADRID)[5] (2012–PRESENT)**

3.1.4.1 Project Synthesis

The general aim of this initiative was the revitalization of the public ways, droveways, and banks of the Jarama River in the municipality of Ciempozuelos (Fig. 3.19). We sought to develop a detailed understanding of the history of the area up to the present day as a basis for imagining a brighter future for these underutilized public-domain lands, and developing proposals to bring them to life once again.

The preliminary stage involved a field-based assessment of the condition of the droveways, public paths, and riverbank in Ciempozuelos and a review of the legal framework applying to these public-domain lands.

The network of droveways of the Iberian *meseta*[6] is a highly complex system, closely intertwined with the territory that now comprises the Madrid region, and forms part of its cultural identity, fauna, traditional architecture, and natural and

FIGURE 3.19

The Jarama River as it passes through Titulcia and Ciempozuelos municipalities (Madrid).

[5]Collaboration with associations Consumo-Gusto (Ciempozuelos) and ARBA.
[6]The *meseta* is a high plateau occupying most of central Iberia and divided into two parts by a mountain range known as the *cordillera central*. The northern part of the *meseta* primarily comprises the region of Castile and Leon and the southern part is occupied by Castile La Mancha. The Madrid region, formerly a northern province of Castile La Mancha, is bounded to its north by the *cordillera central*, and is thus a key meeting point for the many droveways crossing the *meseta*.

agricultural landscapes. Defining clearly both the boundaries of the lands in the public domain and the situations that could endanger them is necessary not just to preserve the public spaces themselves but also to reduce the threat of impacts to adjacent land in private hands.

This project is a long-term collaboration aimed at municipal transformation through participatory processes. Each year it is proposed to undertake new activities of dissemination, public awareness, and local community empowerment.

3.1.4.2 Description of Techniques

This project is ongoing. So far a range of approaches has been employed, including transect walks (Chapter 2, Section 2.1.2.1.4) and participatory mapping (Chapter 2, Section 2.1.2.3.2.2). Future techniques will focus on dissemination and empowerment of civil society.

3.1.5 CASE 3.1.5: COMMON GRAZING AND TRANSHUMANT LIVESTOCK MANAGEMENT IN HERMANDAD DE CAMPOO DE SUSO MUNICIPALITY (CANTABRIA)[7] (2011)

3.1.5.1 Project Synthesis

Hermandad de Campoo de Suso has one of the largest areas of common grazing of any municipality in Spain (Fig. 3.20), and as such provides an interesting example of common management of this kind of resource. Until only a few years ago transhumant pastoralism of Merino sheep through the Híjar mountain pass coexisted with grazing of cows and mares belonging to settlements in Hermandad and neighboring municipalities. The use of common grazing has been regulated by commons rules since at least medieval times, preserving productive resources and protecting them from individual overexploitation. The earliest documented example of these regulations dates from 1589.

In this project we carried out an appraisal of the use of these pastures and the management of livestock in the municipality of Hermandad de Campoo de Suso. The study included an analysis of the training requirements for shepherds and farmers in the locality, and assessed the possibility of setting up a school of pastoralism where existing livestock workers could provide training for future generations of pastoralists.

As a result of this work, our group (OCT) has become a member of the Extensive Livestock Farming and Pastoralism Platform (www.ganaderiaextensiva.org/tag/plataformapor-la-ganaderia-extensiva-y-el-pastoralismo/), and we continue to collaborate closely with the association Trashumancia y Naturaleza (www.pastos.es) in the promotion of transhumance and continuation of the pastoral way of life.

---

[7]Funded by Asociación País Románico.

FIGURE 3.20

High grasslands, Hermandaz de Campoo de Suso municipality (Cantabria).

3.1.5.2 Description of Techniques

The techniques most commonly used in this project were questionnaire surveys (Chapter 2, Section 2.1.2.1.2.1) and discourse analysis (Chapter 2, Section 2.1.2.2.6).

3.1.6 CASE 3.1.6: PROJECT ASSESSMENT: FOOD SOVEREIGNTY CONSOLIDATION IN COUMBACARA, SENEGAL[8] (2015)

3.1.6.1 Project Synthesis

Under the overall aim of poverty reduction in Coumbacara rural community, the project defined the following objectives.

1. Development of a plan for the qualitative and quantitative improvement of production to ensure food security.
2. Development of a women's training plan focused on acquisition of knowledge and skills and strengthening and enhancement of local organization and networking.
3. Empowerment of women through income diversification and facilitation of their tasks in the agrifood sector.

OCT members, in the role of external evaluators, carried out the evaluation of the project, which had the following specific objectives.

---

[8]Agencia Andaluza de Cooperación Internacional al Desarrollo. Project coordination: Paz y Dignidad association. Project implementation: local partner 7° and National Association for Adult Literacy and Training.

1. To draw conclusions and recommendations for local institutions, funders, and development agencies working in Senegal to improve the quality of their operations, redefining their strategies, methodologies, actions, and internal organization.
2. To assess the extent to which actions, results, and objectives had been reached through a comparative analysis of designed operations (program theory) and actual implementation (implementation theory).

Stakeholders involved in the project implementation were also actively involved in the external evaluation. Both the local partner and Paz con Dignidad worked together with the evaluation team in the design, fieldwork preparation, validation of tools, etc. This resulted in a learning process for participating stakeholders, allowing methodologies and tools to be exchanged, and the findings, analysis, value judgments, and recommendations to be clearly linked. The results of the evaluation emerge from the stakeholders, with the evaluation team playing the role of catalyst and facilitator.

The information needed for the evaluation was collected in situ, through group working in participatory workshops and techniques carried out by local actors on an individual basis.

3.1.6.2 Description of Techniques
3.1.6.2.1 Participatory Mapping (Chapter 2, Section 2.1.2.3.2.2)
The participatory mapping approach described here is usually applied in the context of participatory rural appraisal (Chambers, 1997), under a methodology called farmer participatory research (Farrington & Martin, 1988). Although the technique of participatory mapping is discussed in Chapter 2 under "analysis techniques," here we use itwithin the framework of a participatory rural evaluation process. The technique is developed through focus groups and involves the participatory development of a map defining the local population's vision of the space and the use of resources within it, locating the principal relevant information. Participants depicted key elements like the horticultural area, dykes, mills, and storehouses using simple cartographic symbols (Fig. 3.21). They then linked these to their outputs (products, by-products, waste, etc.) on a separate wallchart using arrows and connection lines. Finally, the participants represented the destination of these outputs (home for self-consumption, market, etc.) on the wallchart by means of arrows. Throughout this process, facilitators noted the ideas expressed by participants using a range of pictographic symbols and images in different colors and materials that everyone could easily understand, trying at the same time to keep to a minimum the amount of information contained on a single chart to avoid confusion.

For the evaluation of the horticultural area, the workshop participants were women beneficiaries from Sare Sambel and Thidelly (Fig. 3.22). The technique was complemented with a catchment diagram to identify problems and potentialities linked to water access in the horticultural area. Participatory approaches were used to decide which stakeholders were responsible for different tasks, who could access

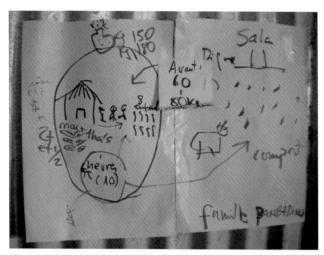

FIGURE 3.21

Depiction of one of the farms from the participatory mapping activity.

the products of family work, and how such decisions should be made. This led to a deeper appreciation of gender roles within the family and within the project.

To investigate the use of resources from the construction of dykes, the workshop participants were the beneficiaries of the Bambadinka dyke. In the same way as the previous example, a catchment diagram was used to help identify the drainage pattern and the sub-catchments. This provided the starting point for discussion of environmental interactions in the area of influence of the dyke.

FIGURE 3.22

Beneficiary groups working on the evaluation of the horticultural area.

Similar workshops were held with the beneficiaries of the two mills in Diala-coumbi and Kumbañako, and the beneficiaries of the storehouse in Dialacoumbi.

The graphical representation process enabled people with different levels of education and skills to participate fully, facilitating the empowerment of the communities in the evaluation process. It also helped participants to structure their knowledge and work toward consensus.

3.1.7 CASE 3.1.7: CATTLE DROVEWAYS, PATHWAYS TO HISTORY. LANDSCAPE AND HERITAGE ALONG THE DROVEWAYS NETWORK.[9] (2013)

3.1.7.1 Project Synthesis

Both droveways and other cultural heritage sites enjoy a high level of protection under Spanish national and regional law. Raising the awareness of the population about their value and meaning are key issues for their preservation, thus a joint approach can optimize the use of public resources to this end. However, at present, droveways and other cultural heritage sites are not managed in an integrated way. The relationship between the droveways and the other sites is not well understood, and there is little publicly accessible information available that would allow them to be explored together.

This project therefore proposes to link the wide network of droveways with other cultural heritage elements like archaeological sites, areas of ethnographic interest, etc., often located in the same territory, and undertake a broad strategy of dissemination and public awareness.

The general objectives of the project are as follows.

1. Study the cultural heritage adjacent to droveways in the Madrid region.
2. Develop a participatory mapping project with the aim of collecting geographical information and local knowledge through crowdsourcing approaches.
3. Identify and design cultural interest routes for local communities to develop guided itineraries and promote their own sustainable tourism activities.
4. Disseminate the information through multimedia tools and mobile applications (Fig. 3.23).

3.1.7.2 Description of Techniques

This project is ongoing, as we have not been able to secure funding. Some limited literature review has been carried out and we have visited some of the key archaeological sites of interest. The main task completed is described as follows.

---

[9]In collaboration with Área Arqueología cooperative.

FIGURE 3.23

10.5 km route in Navalagamella (Madrid). Virtual flight over several droveways, using publicly available aerial photographs.

3.1.7.2.1 Digital Visualization and Multimedia (Chapter 2, Section 2.1.2.3.4)

A virtual tour of droveways and cultural heritage sites in the Navalagamella area was developed using fly-through animation techniques and publicly available aerial photographs. The work was undertaken by Rubén Suárez Iglesias, a student on the geographical information technologies master's degree course of the University of Alcalá, Madrid, as part of the practical experience component of his degree. The visualization can be found at www.youtube.com/watch?v=7A6z62UTxo8.

3.1.8 CASE 3.1.8: PUBLIC PARTICIPATORY GEOGRAPHIC INFORMATION SYSTEMS: ADOLESCENT PERCEPTION OF LANDSCAPE ELEMENTS, VALUES, AND CHANGES IN COLMENAR VIEJO (MADRID)[10] (2016–2017)

3.1.8.1 Project Synthesis

Sustainable Futures for Europe's Heritage in Cultural Landscapes (HERCULES— see http://www.hercules-landscapes.eu/) is a transdisciplinary European research project whose aim is to understand the elements, values, and drivers of change in Europe's cultural landscapes and produce knowledge to ensure their protection, planning, and management. To achieve this, six regional landscape cases were chosen, representing the environmental gradients and land uses in Europe as a whole. One of these was the Madrid municipality of Colmenar Viejo, in the foothills of the Guadarrama mountain range.

[10]On behalf of the University of Hohenheim, Germany, under the HERCULES project, funded by the Seventh Framework Programme of the European Union. Task coordinated by Claudia Bieling.

In this project we took part in researching the perception of adolescents in the municipality of Colmenar Viejo about the landscape in their locality using public participatory geographic information systems (PPGIS), an approach that is becoming increasingly popular for territorial decision-making.

The work had the following aims.

1. To identify the tangible and intangible landscape elements.
2. To identify the associated emotional meanings.
3. To identify the landscape services associated.
4. To identify the landscape elements or uses that are most widely perceived to have changed.
5. To contribute to the ongoing methodological discussion about the use of PPGIS in adolescent populations.

3.1.8.2 Description of Techniques

3.1.8.2.1 Questionnaire Surveys (Chapter 2, Section 2.1.2.1.2.1) and Public Participation Geographical Information Systems Approaches (Chapter 2, Section 2.1.2.3.2)

The method used an online questionnaire survey developed for the HERCULES project and implemented in Maptionnaire software (Fig. 3.24). The survey involved a series of questions designed to elicit adolescents' perception of services and values that the landscape may give them, such as a place to play sport or appreciate the tranquility of nature. They were also asked questions relating to changes that they might have perceived, such as urban expansion or abandonment of agricultural land.

At present we are working on analysis of the survey data using a range of statistical and geographical analysis methods.

FIGURE 3.24

Screen capture from the online survey completed by adolescent participants.

3.2 THEME 3.2: BETWEEN CITY AND COUNTRY: BUILDING MORE RESILIENT RURAL—URBAN RELATIONS

INTRODUCTION

Rural areas are often seen as the opposite or inverse of urban areas; the junior partner in an unequal relationship. Cities are viewed as dynamic, vibrant, innovative places, while the countryside is frequently perceived as out of step with the times, the place to go for leisure or retirement, where nothing of any importance or interest can happen. Sometimes (far worse), the country is thought of as a kind of city-in-waiting, an empty space ready to be urbanized just as soon as conditions are right.[11] Yet rural and urban areas are much more diverse and interconnected than popular myth allows. The relationship between them is highly dynamic and deeply dependent on the movement of people and, increasingly, the movement of goods and services. Rural and urban are not opposites in any real sense, but complementary elements of the same system. Cities may be important centers of knowledge exchange, commerce, and employment, but the country is where most of the food is produced, where raw materials are found (stone, clay, timber etc.), and where nature's services provide for and ensure our well-being. We cannot have one without the other. Yet in recent times, as cities and their surrounding rural regions have become ever more tightly coupled to global systems, they have become more disconnected from each other, and societies are as a result increasingly vulnerable to catastrophic cascade effects from global crises (Homer-Dixon et al., 2015). Cities have surrendered their identities as territorial capitals and centers of local exchange and become service hubs for global businesses. Rural areas are increasingly threatened in a range of ways, by abandonment of traditional agriculture, by degradation from vast mono-cropping programs or urban sprawl, by depopulation, by cultural decline, by unemployment and low wages, by loss of services, and ultimately by loss of identity. These changes have come about for a number of reasons:

Firstly, urban areas are now vastly more populous and demographically dominant than they once were. Recent decades have seen rapid urban growth worldwide, especially in developing countries,[12] but also in Europe, the United States, and Australia. This pattern is part of a global transition in population distribution away from rural areas to urban centers. Today more than 50% of people worldwide

---

[11]This view was cogently expressed by Esperanza Aguirre, president of the Madrid region (2003—2012) in a television interview of 2003: "Así, una vez que se sepa que no hay nada que proteger…se podrá construir. [In this way, once we know that there is nothing to protect…we will be able to build.]" (TeleMadrid, 2003). The implication is not only that protected areas are an inconvenience to be overcome, but that land lacking protection is automatically appropriate for urban use.

[12]We accept, and are sensitive to, the criticism that the use of terms like "developed" and "developing" implies unfounded value judgments, yet we have been unable to find better alternatives. In our defense, as readers will hopefully recognize from the text, we do not consider the presence or absence of what has conventionally been called "development" to be a useful yardstick for social progress.

live in a city (UN, 2005), and this figure is expected to grow in the future as the global population increases. In Spain, where the case studies in this section are located, impoverishment of the countryside and industrial growth in urban areas during the dictatorship of Francisco Franco (1939−1975), particularly after 1960, led to a wave of migration to the major cities and industrial areas, notably to Madrid, Barcelona, and Bilbao (Trimiño, 2013). Most of the migrants came from the rural agricultural regions of Castille-Leon, Castille-La Mancha, Andalusia, and Extremadura. This led to a major decline in the rural population, with an estimated 1 m agricultural workers leaving the countryside between 1960 and 1972 (Trimiño, 2013, p. 23). The massive influx of population to the cities led to rapid and widespread urbanization. The growing economy, the tourism boom, and the adoption of increasingly affluent lifestyles has meant that significant urban expansion continued right up to the global financial crisis of 2007. The hunger for building land, especially after 2000, when the housing market was at its most overheated, led to indiscriminate urban development almost everywhere, including in natural areas and on prime-quality agricultural land (Hewitt & Escobar, 2011; Hewitt & Hernandez-Jimenez, 2010; Romero, Jiménez, & Villoria, 2012).

Another important reason for transformation of the relationship between city and country is globalization. Great improvements in speed, efficiency, and availability of transport, powered by cheap and abundant oil, mean that an enormous range of goods and services can be distributed and sold far from their place of production. Though international trade and exchange are probably at least as old as agriculture itself, not until the 20th century did it become commonplace for tomatoes, for example, to be simultaneously grown in the Netherlands for sale in Italy and grown in Italy for consumption in the Netherlands (de Pablo Valenciano & Mesa, 2004; CBI Market Intelligence, 2015), or for oranges from Argentina or Chile to be transported to market in the northern hemisphere summer (Friedland, 1994). And by taking advantage of differentials in living costs between countries and the wide variation in currency prices, it is now also common for a single product to travel backwards and forwards several times between three or more continents from the place where it is grown to its point of sale. Clearly, whatever the benefits that may be argued in terms of generating employment in developing countries, driving down food prices to the consumer, and ensuring a diversity of produce to consumers (mostly in the developed world), the environmental impact of such a model is enormous, in the main due to the vast distances that produce must travel before it arrives at the market. As a consequence of the large capital outlay necessary to get established, as well as the complicated logistical arrangements, this model also tends to favor large multinationals, not local exporters or small farmers, and is dependent on many intermediaries, each of whom takes a cut. For the consumer to enjoy low prices, the producer's profit is often squeezed by other actors, such as wholesalers and supermarkets, with positions of power in the value chain. Another very notable consequence of this model is that regions, and sometimes whole countries, may lose their markets entirely in response to emerging competition or agricultural policy decisions. This not infrequently results in the dramatic decline of the local agricultural economy, and in some cases wholesale abandonment of productive farmland, with its accompanying loss of traditional skills, knowledge, and cultural practices. In Tenerife, for example, the once vibrant agricultural sector is in the doldrums, with an estimated 60% of farmland abandoned, thanks to the loss of some of its major markets due to competition elsewhere (Gobierno de Canarias, 2015).

Finally, the increasing mechanization and industrialization of agriculture, which has been both a driver and a consequence of globalization, has played a strong role in transforming agricultural landscapes and societies. Crop standardization and the near-universal global adoption of a few commercial varieties have undermined the territorial identity of many rural areas. The coevolution of society and territory that has lasted for millennia in agrarian spaces has all but disappeared. Intensification has transformed picturesque rural landscapes with vibrant communities into barren wastelands of monotonous uniformity dependent on seasonal flows of (low-paid) migrant labor; witness, for example, the famous *mar de plástico* (sea of plastic) in the horticultural areas of the Spanish province of Almería.

Although this industrialized model of agriculture retains its dominance in most parts of the world, awareness of its negative impacts is growing and it is now being widely called into question. The Fair Trade movement, for example, strives to address the exploitation of producers in developing countries through direct marketing to consumers and certification and labeling initiatives (Raynolds, Murray, & Wilkinson, 2007). Agroecology offers a practical, evidence-based response to these challenges, seeking to improve the sustainability of food systems by applying ecological and social science approaches (Méndez, Bacon, & Cohen, 2013). Parallel initiatives are emerging which aim to increase awareness of the value of rural areas and help them recover their territorial identity. Farmers' markets, community-supported agriculture initiatives (CSAs; Brown & Miller, 2008), and organic box schemes (Brown, Dury, & Holdsworth, 2009) have emerged as popular alternatives to traditional food networks, though primarily among more affluent consumers—one of the main challenges is extending the involvement of lower-income communities (see, for example, the first case study in this theme, 3.2.1: Ecosocial market gardens). Nonetheless, the growing interest in local produce and landscape associations of food has the potential to strengthen networks of interdependence between rural and urban areas, increasing economic opportunities for rural communities and improving the quality of urban life.

Rethinking the relationship of the city with its rural surroundings has become a central issue of contemporary urban debate, with many social sectors (farmers, consumers, civil society, academics, and environmentalists) calling for a more symbiotic and harmonious relationship between urban and rural areas. Interest is growing in alternative planning models, like the *bioregion* (Thayer, 2003), a naturally bounded region of shared ecological and landscape characteristics, encouraging people to build strong links to their environment and a sense of responsibility for its stewardship. Adopting a bioregional approach does not necessarily require major structural or legislative changes. In many parts of Europe, for example, the smallest administrative areas (in most parts of Spain this is the municipality) frequently already form part of a recognizable (ancient) territorial unit. These are often essentially self-governing with minimal resources, and local associations and community groups with a strong sense of local identity may already play a major role in civic life. In such cases progress can be made by formalizing existing tendencies and practices[13] and working with communities to build on the existing feelings of place-based identity. As the process develops, neighboring municipalities can join together to promote the idea of a shared regional identity.

One key area of recent focus has been peri-urban areas, the transitional zones, outskirts, or hinterland of a city, typically an unplanned jumble of industrial units, infrastructure, waste dumps, and abandoned farmland. In general, peri-urban areas have been neglected by regional planners and citizens alike. This is a clear missed opportunity—these spaces have great potential and versatility for addressing urban provision of food and other ecological services, decreasing vulnerability to global shocks, and enhancing regional resilience. Restoring the broken links between city and country necessarily involves rethinking the role of farmland on the periphery of the city (Hernandez-Jimenez et al., 2009; Montasell, 2006a, 2006b; Moratalla, 2015; Ochoa & Moratalla, 2015; Simón et al., 2014). As Morán Alonso (2015) observed, the role of the food system has been generally overlooked in regional territorial planning. In fact, a radical structural reorganization of all aspects of the food system (production, processing, distribution, and waste) is key to rebuilding the relationship between urban and rural areas.

In peri-urban areas, even where the landscape is seriously degraded or impacted by infrastructure, services, or urban expansion, local communities may retain a strong sense of place-based identity. In many such areas a wide range of initiatives are already being implemented; some of them are described in this theme. They include land banks, agrarian parks (Moratalla, 2015), allotments and market gardens (for both recreational and commercial purposes), training projects oriented to local land-based entrepreneurship, and other schemes to promote the recovery of abandoned or underutilized land. In some cases these initiatives try to bring about social transformation of the territory through agroecological transition processes at the municipal or regional scales. In other cases work is more local in scale, directed to launching community-based food production activities and building and strengthening alternative means of commercialization. It is important to note that these kinds of initiatives imply changes in the way the territory is managed and the land is used. This is a rich source of conflict between stakeholders belonging to different interest groups and sectors of society. Consequently, participatory approaches that provide spaces for dialogue and consensus are essential to facilitate the transition to more sustainable and resilient planning systems in the long term.

The projects described in *Theme 3.2: Between city and country: building more resilient rural-urban relations* include both civil society and academic research initiatives. Their common aim is locating food and agriculture as a central axis for the

---

[13]E.g., rights to firewood or grazing in return for assistance with woodland or pasture management, or communal rights and responsibilities as part of an association of water users. In Spain these ancient forms of community stewardship still exist in many rural areas.

FIGURE 3.25

Ecosocial vegetable gardens in Azuqueca de Henares.

reconstruction of rural—urban linkages and relationships, from both a geographical and a social perspective.

Civil society initiatives include projects addressing social and leisure activities, self-sufficiency, and commercial agriculture, as well as training support for agroecological transition and alternative means of distribution (short food supply chains). These projects are mostly small in scale and generally combine a few rapidly executed actions with network-building activities. The projects are described together as a single case as follows.

3.2.1: Working for an Agroecological Transition

- Ecosocial vegetable gardens in Azuqueca de Henares (Fig. 3.25).
- Community-supported agriculture in Zarzalejo transition town.
- The "Somos Vega" platform.
- The Madrid Food Sovereignty Initiative (currently part of the Madrid Agroecology Platform)
- The Ecos del Tajo transnational and international cooperation project.
- Agroecology in the Sierra Oeste.

Academic projects are generally broader in scope, with fewer specific actions at the community level. In general these projects search for ways to develop more representative forms of land governance through engagement with citizens as active decision-makers. All the projects look to increase resilience against global shocks like the 2007 financial crisis by developing new tools and approaches to land planning which put agriculture and food systems at the heart of inclusive multifunctional territories. These approaches explicitly define the territory as a common resource, not as a consumer product (Pacheco & Hewitt, 2010), and emphasize the need for land use to be decided by sustainability criteria and the needs of citizens rather than the "invisible hand" of the market. The following specific case study projects are described.

3.2.2 European Cooperation in Science and Technology (COST) Action Urban Agriculture in Europe TD1106 (2012–2016), a research network led by Aachen University (Germany) and Milan Polytechnic University (Italy).

3.2.3 Peri-urban Agrarian Ecosystems in Spatial Planning (PAEc-SP) (2012–2014), a collaboration with the Department of Urbanism and Spatial Planning, School of Architecture and Research Group on Ecology and Landscape, Polytechnic University of Madrid (Spain).

3.2.4 Environmental perception of farmers in Las Vegas County, Madrid (2012–2013), in collaboration with the Madrid Institute for Rural, Food, and Farming Development (IMIDRA).

3.2.5 Integrative Systems and Boundary Problems (ISBP) (2006–2009), in collaboration with the Department of Rural Planning and Projects, School of Agricultural Engineering, Polytechnic University of Madrid (Spain), and Newcastle University (United Kingdom).

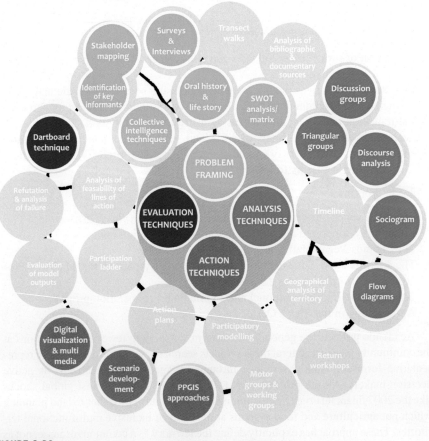

FIGURE 3.26

General methodological framework for Theme 3.2: Between city and country. The techniques we have most frequently used in this section are highlighted.

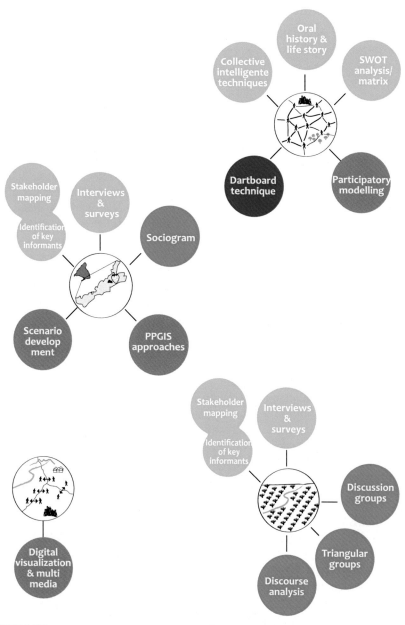

FIGURE 3.27

Diagram showing the relationship of the techniques used, following the color coding in Fig. 2.2 (Chapter 2), to the projects described in this theme. Each project is given a symbol, which appears in the top right-hand corner of each factsheet to help the reader identify the project.

GENERAL APPROACH AND STRUCTURE

In this section we present the principal tools and approaches that appear in the case studies described in *Theme 3.2: Between city and country.* In Fig. 3.26 we highlight the tools that are most relevant and most frequently applied to this theme. These techniques are described in detail in Chapter 2. We then present the case study projects listed above to show how these techniques have been applied in practice in a diverse range of situations.

The techniques described in this section have two main objectives: empowering social actors through the development of actions at the local and regional scales, and building groups and networks for knowledge transfer among actors. The first objective is addressed by appraisal techniques like *semi-structured interviews* (Chapter 2, Section 2.1.2.1.2.2) and analysis techniques like *discussion groups* (Chapter 2, Section 2.1.2.2.3), *triangular groups* (Chapter 2, Section 2.1.2.2.4), and *flow diagrams* (Chapter 2, Section 2.1.2.2.5). The second objective is developed through analysis techniques like *sociograms* (Chapter 2, Section 2.1.2.2.2) and *discourse analysis* (Chapter 2, Section 2.1.2.2.6), and action techniques like *participatory mapping* (Chapter 2, Section 2.1.2.3.2.2). Fig. 3.26 shows all the techniques used in this theme; Fig. 3.27 shows how each technique relates to a specific project.

DESCRIPTION OF PROJECTS AND TECHNIQUES USED

As for the previous theme, a synthetic review of each project included in *Theme 3.2: Between city and country* is presented in the following pages, with a brief description of the techniques employed in each case. This review is presented as a factsheet containing the basic information about the project: its geographical location, context, funding body, main objectives, and a synthetic description of the work carried out. As in the previous theme, we remind the reader that we do not provide an exhaustive description of each and every technique used in a given project, but rather we have selected what we regard as the most important or illustrative techniques relevant to the theme for detailed treatment.

3.2.1   CASE 3.2.1: WORKING FOR AN AGROECOLOGICAL TRANSITION (2009–PRESENT): ECOSOCIAL VEGETABLE GARDENS IN AZUQUECA DE HENARES, ZARZALEJO TRANSITION TOWN, SOMOS VEGA PLATFORM, MADRID FOOD SOVEREIGNTY INITIATIVE, ECOS DEL TAJO TRANSNATIONAL AND INTERNATIONAL COOPERATION PROJECT, AGROECOLOGY IN THE SIERRA OESTE

3.2.1.1 *Project Synthesis*

This broad case study groups together several rural–urban linkages and transition spaces projects. These initiatives aim to recover agricultural areas for productive economic activity by strengthening and emphasizing their multifunctionality and biodiversity. Land recovery initiatives go hand in hand with social and cultural revitalization, and help to replicate similar initiatives in the territory.

The common objectives of these initiatives are listed below.

- Participatory territorial assessments (farm and agriculture sectors, commons, cultural heritage, etc.).
- Local and traditional knowledge collection.
- Seed banks and native species networks for preservation, reproduction, and reforestation.
- Community composting.
- Community gardens and municipal vegetable gardens directed to new supply chains.
- Recovery of commons (agroforestry systems, public-domain waterways and water resources, droveways, etc.) for agricultural and pastoral activities and the enjoyment and benefit of the population as a whole.
- Dissemination to increase public awareness of the value of agrarian spaces in urban–rural transition zones, as well as development of actions aimed to reconnect agriculture, environment, and society.
- Promoting the involvement of local communities in land planning and management.
- Transfering the necessary conceptual and methodological tools to local populations to allow them collectively to continue the local development processes initiated.
- Building links between different stakeholders in *bioregions* and historic counties.
- Creation of a land school/rural university (advising on agroecological farming and cultural and natural valorization).
- Launching a land stewardship network and facilitating stewardship agreements.

In parallel with the development of these actions we participated in the Madrid Food Sovereignty Initiative, a meeting platform and lobby group for collectives involved in agroecological production and consumption, established in 2011. Following a participatory process to define strategies, the Madrid Agroecology Platform was formed from the agroecological movement in Madrid.

The individual projects are described below.

3.2.1.1.1 Ecosocial Vegetable Gardens in Azuqueca de Henares[14]

This project began in 2014 and is currently ongoing. Azuqueca de Henares is a traditionally agricultural municipality, although in recent decades the agricultural sector has come under strong pressure from industry and construction. As a result of the

---

[14]Full title: Ecosocial vegetable gardens, eco-entrepreneurship school and land bank in Azuqueca de Henares, social and environmental integration initiative. Led by OCT and the Guadalajara Organic Farmers' Association (Güecológico), and funded by Azuqueca Town Council.

FIGURE 3.28

Ecosocial vegetable gardens in Azuqueca de Henares.

financial crisis, economic activity has declined and unemployment has risen, leading to a strong increase in the number of people at risk of social exclusion. In this context, new public policies have been developed to try to unlock the potential of agroecology to address social demands and help create a more sustainable local economy.[15] On this basis, a strategy for creating a system of agroecological areas (known by the Spanish acronym SEA) was developed, with three basic pillars: economic, social, and environmental. A small organic farmers' collective has been established, composed of socially vulnerable inhabitants of the municipality, with the aim of managing an area of land intended for self-production (Fig. 3.28). The intention is to support participants interested in developing small-scale organic agriculture as a professional activity to supplement their income and increase their self-confidence and social inclusion.

The following main actions have been developed (see http://azuquecologicos. blogspot.com.es/p/proyecto-huertos-ecosociales-de.html for further information in Spanish).

- Supporting and guiding users of the ecosocial vegetable gardens.
- Providing training in agroecological production and distribution systems through an eco-entrepreneurship school.

---

[15]The Terrae Municipalities Network was launched in 2012; it has a highly practical focus on sustainable development through agroecology and food sovereignty as a driver of social transformation and local development linked to land stewardship. Azuqueca de Henares joined the network in 2014, beginning the ecosocial vegetable gardens project located in Quebradilla Park. The project described here is led by OCT and Güecológico, working with Azuqueca Council for the implementation and monitoring of an eco-entrepreneurship school and a land bank.

- Designing and maintaining a land bank, with the aim of establishing an agro-ecological municipal park.

Other participatory activities proposed for supporting this project are the following.

- Promoting the integration of the local community in the planning and management of municipal lands as a means to reinforce their sense of belonging and of shared responsibility.
- Establishing participatory monitoring mechanisms to address problems and build on success stories in future years.

This project is ongoing (provided local-level funding can be maintained), allowing a continuous process of transformation and community engagement.

3.2.1.1.2 Zarzalejo Transition Town

The Zarzalejo transition town initiative (http://zarzalejoentransicion.blogspot.com.es/) was launched 2011 as a community response to the search for new models of living. The initiative, inspired by the global "transition towns" movement (https://transitionnetwork.org/), is entirely managed by local residents. It aims to address global problems through specific local actions like collective transport groups, renewable energy development, and community-supported agriculture (CSA) schemes (CSA Zarzalejo is a socially responsible agriculture and agroecological community garden).

3.2.1.1.3 Somos Vega Platform

Somos Vega is a community project initiated in 2013 that links several municipalities in the proximity of the Jarama River, located in the south of the Madrid region. It aims to recover public-domain areas for farming and agriculture activities as well as for the enjoyment of the local population; to involve local, environmental, and cultural initiatives in the revalorization of peri-urban areas; to recover native species; and to decrease environmental degredation in the areas adjacent to the Jarama River.

3.2.1.1.4 Madrid Food Sovereignty Initiative

Launched in 2009 by the agroecological movement in Madrid, this initiative led to the establishment of the Madrid Agroecology Platform in 2016, responsible for the coordination of the diverse food sovereignty and agroecology initiatives located in the Madrid bioregion. Our group (OCT) has participated since 2009, working on two initiatives related to the development of alternatives to existing market-based approaches from the perspective of de-growth (Martínez-Alier, Pascual, Vivien, & Zaccai, 2010) and food sovereignty (Altieri, 2009).

3.2.1.1.5 Ecos del Tajo Transnational and International Cooperation Project

This four-year project (2009–2012) was based in the Tajo River catchment in Spain and Portugal. Seven local action groups[16] from Cáceres, Madrid, and Guadalajara took part. The Comarca de Trasierra–Tierras de Granadilla Development Association was the coordinating partner responsible for project implementation. The specific objectives of the project were as follows.

- Fostering agroecological production linkages in the Tajo River catchment bioregion and between sub-catchment areas.
- Developing land stewardship programmes.
- Creating seed preservation and exchange networks.
- Providing training and advice on agroecological production.

OCT worked with the agroecological development group Red Calea in both the Sierra Oeste area of the Madrid region and the Molina de Aragón–Alto Tajo area in Guadalajara province. The collaboration involved providing training and advice on agroecological production, and holding participatory workshops for improving organic food commercialization through short supply chains.

3.2.1.1.6 Agroecology in the Sierra Oeste[17]

In the Sierra Oeste (Western Uplands), an area of upland forest and pasture in the west of the Madrid region (Fig. 3.29), a variety of bottom-up initiatives have come together in recent years, coinciding with the systemic crisis (economic, environmental, social) which Spain is currently undergoing. These initiatives have in common the desire to explore the agroecological potential of this historic country. In this context, the Sierra Oeste Agroecology Project (see: http://sierraoesteagroecologica.org/) was established to support the transition toward agroecological systems from county to local level. The project defined the following objectives for 2017–2018.

- Preparation of an appraisal of resources and actors in the county involved in agroecology.
- Creation of the Sierra Oeste Agroecological Association as a forum for coordination of entities and individuals who are interested in working on this theme.
- Setting up a pilot agrocomposting project in Zarzalejo and Fresnedillas de la Oliva municipalities.

---

[16]Local action groups or *Grupos de Acción Local* (GAL) in Spanish are "associations or other not-for-profit entities whose remit is to manage local development strategies in specific rural regions in line with European Union regulations" (http://www.redruralnacional.es/leader/grupos-de-accion-local).

[17]Project coordinated by OCT, Germinando, and Zarzalejo Town Council, financed by the Nina Carasso Foundation. Also supported by Fresnedillas de la Oliva Town Council, Madrid Agroecológico, and Zarzalejo en Transición (2017–ongoing).

FIGURE 3.29

The River Cofio as it passes through the municipality of Valdemaqueda in the Sierra Oeste.

Photograph: Alicia López Rodríguez.

- Raising awareness among the local population about agroecology and the existing resources for its development in the county.

The project began with workshops in two municipalities of the Sierra Oeste, Zarzalejo and Chapinería. A diverse range of actors including farmers, land planners, civil society organizations, and politicians from all over the Sierra Oeste were invited to participate and share their knowledge about local resources and initiatives. The first workshop was held on March 25, 2017 at the cultural center in the village of Zarzalejo Estación. Though the first workshop was well attended, with 30 participants, local administrators and development agents were poorly represented. This was unfortunate, since these actors occupy a key role as facilitators for local initiatives. For this reason, at the second workshop, held on April 24, 2017, a big effort was made to get these actors to attend, going to the trouble of choosing a local government site, the environmental education center in Chapinería (part of the official network of environmental education centers of Madrid) to host the workshop. Unfortunately, the hoped-for participation of these key local functionaries was not achieved, and the new location presumably put off some other participants. In the event only 10 stakeholders attended.

The workshops, entitled "Workshop for participatory mapping of actors and agroecological resources in the Sierra Oeste," followed the same structure in each case and had the following aims.

- To present and disseminate information about the initiative.
- To begin to establish a network of parties and individuals interested in agroecology.
- To identify the actors and resources for agroecology in the Sierra Oeste.

3.2.1.2 Description of Techniques

3.2.1.2.1 Ecosocial Vegetable Gardens in Azuqueca de Henares

3.2.1.2.1.1 Oral History and Life Stories (Chapter 2, Section 2.1.2.1.5)—a Collective Seasonal Crop Calendar

A key aspect of agroecological initiatives at the local scale is the recovery of local customs and techniques that, in many cases, exist only in the memories of local people. For the Azuqueca ecosocial vegetable gardens project we worked to build a collective seasonal crop calendar using traditional ecological knowledge held by local inhabitants. This is an unusual and interesting use of oral historical or life knowledge that we feel is worth reproducing here.

This exercise was carried out within the framework of the eco-entrepreneurship school initiative. We brought together senior farmers from the farmers' association of the municipality and new farmers from the eco-entrepreneurship school. The aim was to represent a calendar of productive activities using traditional varieties cultivated in Azuqueca de Henares for more than 60 years. Over the course of a 3 h session participants were able to remember the cultivated varieties of dry and irrigated cereals and those grown in smallholdings for feeding both the family and its cattle. They also commented on traditional techniques of utilizing agricultural remains for livestock feed, as well as the bioproducts of livestock farming for organic fertilization. Some customs were also recovered relating to community organization, temporary migrations for agricultural work, pest and disease control, and traditional agricultural and livestock implements. The information recovered was documented in detail for future use.

3.2.1.2.1.2 The Dartboard Technique (Chapter 2, Section 2.1.2.4.3)

As noted in Chapter 2 (Section 2.1.2.4, Evaluation techniques), project evaluation should if possible be carried out as an ongoing project activity, not left until the end. In the case of Azuqueca, participatory evaluation is essential to demonstrate the project's usefulness for each new season of work and allow continuous improvement of the vegetable garden users' experience. Adequate and frequent monitoring and evaluation are essential for the ongoing development of this initiative.

The specific objectives of the evaluation in this case were as follows.

- To estimate the success of the project on the basis of the participants' experience.
- To strengthen collective leadership, essential in a project whose ultimate aim is educate and transform society.
- To help take decisions about follow-up activities (project legacy), growth, and changes over time and to finalize strategies.

The evaluation was carried out using the dartboard technique. Participants were asked how to evaluate to what extent the course had met their expectations in the following areas:

- quality of produce obtained
- pest and disease incidences
- adoption of new technology

- adaptation to alternative practices
- cost/benefit
- preparation of biological control treatments
- non-biological control treatments.
- use of common resources (tools).

The dartboard (Fig. 3.30) was displayed on the wallchart and each participant evaluated each objective by drawing a point on the target. The closer a point was located to the center, the higher the score and the more strongly the participant felt his/her expectations had been met. A common reading of the results was then made in plenary session, attempting to group the scores and look for patterns. The reason for the different scores was explored and debated as a group to understand participants' views in more detail and see how poorer scores could be improved.

Though the technique generated very useful and informative feedback, we propose the following improvements for future years.

- Ask participants at the beginning of the project what objectives they wished to achieve by the end of the project.
- Use these objectives directly as the evaluation criteria for the dartboard.

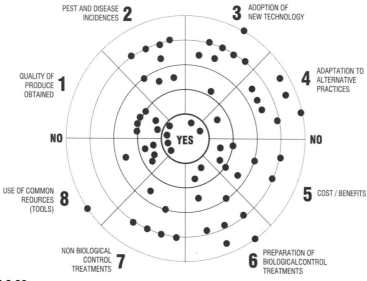

FIGURE 3.30

Participatory evaluation dartboard for the Azuqueca ecosocial vegetable gardens project.

3.2.1.2.2 Ecos del Tajo Transnational and International Cooperation Project

3.2.1.2.2.1 SWOT Matrix (Chapter 2, Section 2.1.2.1.7) and Problem–Action Matrix (Chapter 2, Section 2.1.2.2.5.1)

During the first working session carried out with farmers from the Molina de Aragón area, Guadalajara, within the framework of the Ecos del Tajo project, the work that would be developed in the area was presented. The aim of this first stage of the project was to undertake an appraisal of organic farming in the area. Two techniques were employed: a SWOT matrix and a problem–action matrix. These techniques were used to elicit participants' opinions and attitudes about the potential of the area and its problems for organic farming. Because of the number of participants in the session, four WGs were established. A flipchart, post-its, and pens were given to each group.

Participants in WGs identified the main factors of importance for organic farming at present. They were asked to classify this information under three headings (technical, organization and legislation, and commercialization), and then organize the information into positives (strengths, opportunities) or negatives (weaknesses, threats) for organic farming. This allowed participants to identify the most important critical points or bottlenecks to be addressed as a priority.

In plenary session, a problem–action matrix was constructed from the information provided by all the groups. A double-entry table was drawn on the flipchart. In the table rows, three levels were defined related to the capability of the group to address the problems (see Chapter 2, Section 2.1.2.2.5.1). In the columns the three dimensions used in the SWOT matrix were placed: technical, organization and legislation, and commercialization. The positive and negative aspects identified by participants during the SWOT activity were then located in the relevant cells of the matrix in the flipchart (Fig. 3.31).

In parallel, participants reflected on how different formal and informal *action sets* (groups of associated actors) could intervene to solve the specific issues identified.

These activities led to the definition of four priorities to work on.

- Issues mainly focused on organic farming training in the "medium space." Farmers mainly asked for direct advice in the field.
- Issues focused around farmers' own organizational aspects in the "near space"— in other words, issues that farmers themselves could address.
- Issues focused on the lack of knowledge among farmers about organic commercialization networks in the "near space," also to be addressed by farmers.
- Issues focused on the lack of knowledge among the population about organic food in the "medium space."

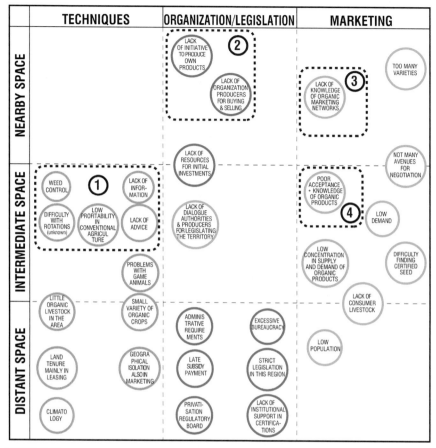

FIGURE 3.31

Problem—action matrix developed for the Ecos del Tajo project.

3.2.1.2.3 Agroecology in the Sierra Oeste

3.2.1.2.3.1 Brainstorming (Chapter 2, Section 2.1.2.1.3.1)

The workshops began with a discussion session in which participants shared their aims and aspirations and identified the main problems and needs in the county, which were added to a wallchart for prioritization. Working individually, participants prioritized the needs displayed on the wallchart using stickers. Participants were given five stickers and instructed to use only one sticker per need. In this way the highest-priority needs were those that had accumulated the greatest number of stickers. The majority of the problems identified related to access to land and lack of processing facilities (like warehouses) for small producers, as well as difficulty in obtaining appropriate certification (e.g., health and safety). The accreditation system was thought to be biased in favor of large businesses. Participants felt that these systems presented numerous obstacles for small enterprises, and as such compared unfavorably with similar systems in France.

3.2.1.2.3.2 Participatory Mapping (Chapter 2, Section 2.1.2.3.2.2)

Identification of actors and resources was carried out by a participatory mapping activity. Participants were divided into four groups, with each group responsible for developing a map under each of the following categories.

1. Businesses and enterprises, e.g., farms, tourism initiatives.
2. Social initiatives or actors.
3. Institutional initiatives or actors.
4. Potential initiatives.

Each group was given a printed paper map of the Sierra Oeste and a sheet of sticky labels of four different colors to identify the locations under each category (Fig. 3.32).

Following the workshops, the four different maps were digitized and combined to produce an online map; this will be made open access to allow participants and other volunteer contributors to update it with new information as the project progresses.

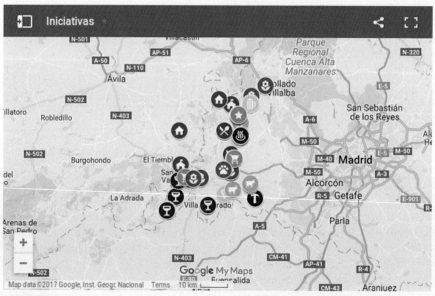

FIGURE 3.32

Participatory map developed by participants in the first workshop. This map was later digitized and uploaded to the web. See http://sierraoesteagroecologica.org/diagnostico/.

3.2.2 CASE 3.2.2: URBAN AGRICULTURE IN EUROPE[18] (2012–2016)

3.2.2.1 Project Synthesis

More than 120 researchers belonging to 61 organizations (research centers, universities, non-governmental organizations (NGOs) and other civil society organizations) from 21 European countries participated in this COST Action. The original aim of the project was to focus European research on urban agriculture (UA) in the framework of future proposals of the common agricultural policy (CAP), and stimulate private and public activities in UA projects and planning. The project was organized in five WGs: WG1 UA definitions and CAP; WG2 UA and governance; WG3 Entrepreneurial models of UA; WG4 Spatial visions of UA; and WG5 UA metabolism.

The project used an innovative methodological approach, integrating bottom-up, top-down, and expert methods, and working in close cooperation with regional key stakeholders. The project has helped to situate UA at the center of the ongoing discussions around the European Union's policy agenda, and to promote a more sustainable and resilient territorial development model.

No specific techniques described in Chapter 2 were employed. Nevertheless we include the description of this project since it is highly relevant to the theme of this chapter. For more information see http://www.urban-agriculture-europe.org/.

3.2.3 CASE 3.2.3: PERI-URBAN AGRARIAN ECOSYSTEMS IN SPATIAL PLANNING[19] (2012–2014)

3.2.3.1 Project Synthesis

peri-urban areas, by virtue of their situation at the urban–rural interface, link social, cultural, economic, and environmental aspects and therefore have great potential to make a positive contribution to achieving sustainable human development. In these areas a diverse range of stakeholders, among them an increasingly mobilized civil society, play a key role. This project aims to integrate Peri-urban agrarian spaces and their associated ecosystems into the urban land system as a means to increase quality of life and social welfare; improve urban and regional resilience; decrease their dependency on external energy; and enhance climate change response and adaptation.

The objectives of the project were as follows.

---

[18]COST Action TD1106 (European Cooperation in Science and Technology project).
[19]National Research Plan 2011, Subprograma Proyectos de Investigación Fundamental no Orientada, Science and Innovation Ministry, Spanish Government. Reference CSO2011-29185. Coordinated by Department of Urbanism of the Architecture School, Madrid Polytechnic University.

- To understand the relationship between ecosystem services (ES) and peri-urban agrarian areas.
- To develop a methodology to identify interactions between urban development patterns and peri-urban agrarian ES.
- To establish a system of indicators and graphical representation methods to understand and communicate the impact on welfare and on promoting biodiversity of the abovementioned relationships.
- To develop a practical tool for assessing agrarian ES oriented toward spatial planning and management, and decision-making on land use.
- To define a methodology that includes the ES approach in the urban planning process.

The project focused on the peri-urban areas of medium-sized inland cities with an agrarian tradition. The town of Aranjuez (southern Madrid region) was one of the case studies.

Four kinds of ES were defined:[20] provisioning services (food production); regulating services (water, climate); supporting services (biodiversity, soil fertility); and cultural services (landscape and identity) (Fig. 3.33). For each service we analyzed stakeholders' responsibilities and relationships, the current state, threats, the expected evolution of the service in question, and existing capacity to ensure its health and maintenance.

3.2.3.2 Description of Techniques

3.2.3.2.1 Stakeholder Mapping (Chapter 2, Section 2.1.2.1.1) and Snowballing (Chapter 2, Section 2.1.2.1.3.2)

The project began with a stakeholder mapping exercise, starting with the researchers' own, very approximate, identification of relevant stakeholders. This first approximation was shared with some of these stakeholders, who suggested others that they considered relevant—an approach known as *snowballing*, as the stakeholder community grows like a snowball as stakeholders identify others to add to the process.

3.2.3.2.2 Surveys and Interviews (Chapter 2, Section 2.1.2.1.2)

Subsequently researchers visited the identified stakeholders and carried out semi-structured interviews (Table 3.2). Interviews had two parts. The first focused on understanding the stakeholders, their roles, and the relationships between them; and identifying new previously undetected significant stakeholders (further enlarging the snowball). In the second part, interviewees were asked to analyze the relationship between ES and different territorial actors. One of the key outcomes of this

---

[20]The definitions used are reported in Simon, Zazo, Moran, & Hernandez-Jimenez (2014), following EME (2011).

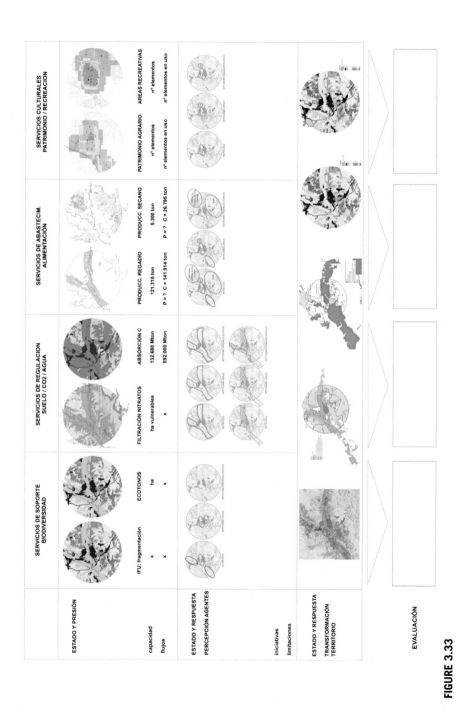

FIGURE 3.33

Ecosystem services assessment exercise for Aranjuez.

Table 3.2 Schedule of Field Visits and Interviews, Peri-urban Agrarian Ecosystems in Spatial Planning Project

Location	Host/Interviewee
11/02/2013	
10:00: La Chimenea research farm	Representative of IMIDRA
12:00: Aranjuez City Hall	Municipal employee from Parks and Gardens Department (president of Cortijo de San Isidro Smallholders Board)
13:30: Aranjuez City Hall	Fundación Paisaje (cultural landscapes foundation).
Afternoon: Cortijo de San Isidro	Young farmer located in Cortijo de San Isidro
17:00: Jaramillo farm	Jaramillo family
Evening: Ecoshop: herbalist and greengrocer	Business owners
18/02/2013	
9:30: Aranjuez City Hall—Urban Planning, Parks, Gardens and Agriculture	City councillor and municipal architect
12:30: Aranjuez City Hall	Aranjuez local development director
25/02/2013	
9:30: Offices of Aranjuez and Las Vegas County local action group (ARACOVE), La Chimena, Real Cortijo de San Isidro, Aranjuez	Former ARACOVE technician and a young entrepreneur
11:00: Aranjuez City Hall	Municipal employee from Environment Department
13:00: Regional government agricultural ministry local offices	Local director, Aranjuez area
05/03/2013	
13:00: Aranjuez market	Representatives of Association of Traders of Aranjuez market (ASCOMAR), municipal market manager, and Secretary of Federation of Madrid Businesses (FEDECAM)
16:00: Tajo River Hydrographic Confederation and Canal de las Aves	Security guard, Tajo Royal aquifer and Canal de las Aves Vice-president of association of irrigators for Canal de las Aves General manager of irrigation area 1 of Tajo River Hydrographic Confederation
Phone call to Ecologístas en Acción environmental association in Aranjuez, visit postponed	

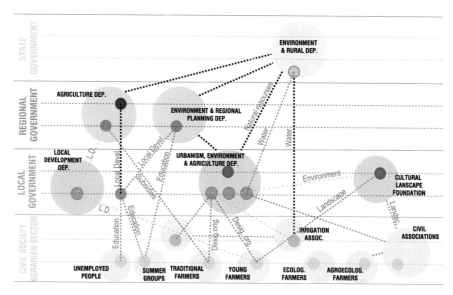

FIGURE 3.34

Sociogram for the peri-urban agrarian ecosystems in spatial planning project.

work was the realization that actors tend to be aware of fewer ES than they actually encounter in their day-to-day activities.[21]

3.2.3.2.3 Sociograms (Chapter 2, Section 2.1.2.2.2)

The sociogram technique was employed in this project as a means of making a rapid appraisal of stakeholders, social networks, and other existing processes in the communities under investigation. This is verified as the participant snowball is expanded (see above), and the role of the different social agents becomes clearer as the process evolves (Fig. 3.34).

3.2.3.2.4 Participatory Mapping (Chapter 2, Section 2.1.2.3.2.2)

The interviews with stakeholders were complemented by participatory mapping activities. Participatory mapping, as a highly visual activity, is an excellent tool for facilitating discussions between actors about the consequences of the loss of agrarian ecosystems under different planning scenarios. It is also very useful as a dissemination tool to help communicate the concept of ES to non-scientific audiences.

In the first of these activities, a workshop was held with a diverse group of stakeholders from the Aranjuez area, including farmers, municipal architects,

---

[21]For detailed discussion of this work see Simón et al. (2014).

and environmental planners. Maps displaying the locations of the four ES types (provisioning, regulating, supporting, and cultural services) in the Aranjuez area were displayed on a whiteboard at the front of the room. The participants were organized into three groups, and each was assigned one of the four ES types; provisioning and regulating services were grouped together, since participants said they found these service types hard to differentiate. The activity was developed in two parts. In the first part participants were asked to indicate, using stickers, locations that they considered important to the provision of these services. For example, for cultural services participants marked historic irrigation infrastructures; under provisioning services, participants located areas with potential for use as market gardens (former market garden areas abandoned or occupied by maize); and for supporting services participants mapped protected areas and reservoirs created from former gravel quarries, like Mar de Ontígola, a well-known local bird habitat. In the second part of the activity, participants debated the difference between the services identified by the research team and the locations the participants considered important, and the value of the ES approach for local planning. Participants recorded the observations arising from the discussion process on post-it notes, which they attached to the relevant ES map in the location relevant to the point they wished to make (Fig. 3.35).

In the second activity, participatory mapping was carried out with local schoolchildren (Fig. 3.36). This activity was undertaken jointly between the PAEc-SP project and another project entitled "Aranjuez, an oasis on the banks

FIGURE 3.35

Participatory mapping working process.

Photo: Surcos Urbanos.

FIGURE 3.36

Children's workshop, "21st century detectives: on the trail of ecosystem services."

Photo: Surcos Urbanos.

of the Tajo River,"[22] with the key aim of disseminating the ES approach more widely through public engagement. Two children's workshops were carried at two Aranjuez primary schools, Santa Teresa (two fifth-grade classes) and Maestro Rodrigo (two fifth-grade classes and one fourth-grade class), organized in coordination with the director of education of the city of Aranjuez. The workshops were entitled "21st century detectives: on the trail of ecosystem services." To accompany the workshops an educational guide was developed for primary school children explaining the ES approach through simple activities. The workshops, held on May 18 and 20, were integrated into the educational activities of each center during school hours. They were conducted jointly by these authors and the association Surcos Urbanos (Urban Furrows), and structured as follows.

- Brief presentation on agricultural ecosystems, the importance to our well-being of the services of these ecosystems, and the value of agriculture in Aranjuez (15 min explanation with audiovisual support).
- Working in groups. Guided by monitors, the students built a model using colored stickers, paper, card, and adhesives, entitled "Agriculture of the 21st century," showing their plan for the spatial organization of two farms in the peri-urban environment of Aranjuez, taking into account the different ES (provisioning, regulating, supporting, and cultural services) (60 min).
- Class discussion and distribution of guides (15 min).

[22]Grant FeCyT-14-9456. Funded by the Spanish Foundation for Science and Technology under the auspices of the Ministry of Economy and Competitiveness. Awarded to Madrid Polytechnic University.

3.2.3.2.5 Scenarios (Chapter 2, Section 2.1.2.3.3)

The second part of the adults' participatory mapping workshop described above involved development of detailed proposals for the future of Aranjuez in 2030 (Fig. 3.37). In groups divided according to sector, participants developed various scenarios for Aranjuez and its hinterland, under the heading "What future do we want for Aranjuez?"

Two questions were posed to help participants orient their work.

1. What changes would be necessary with respect to the distribution of land uses?
2. What implications would these changes have for land planning (with respect to tools, strategies, competencies, and resources)?

The work was undertaken in two groups. The first, denominated the *planners*, included mainly public sector professionals and participants with technical planning responsibilities. The second group, known as *civil society and farmers*, comprised members of civil society organizations, local small and medium-sized enterprises (SMEs), and farmers.

The planners group included two municipal architects, a national government representative for heritage in Aranjuez, the director of security for the Canal de las Aves irrigation canal, a municipal environmental planner, and a representative of Holcim-El Puente, a local quarrying company.

The civil society and farmers group included the president of the association of Aranjuez horticulture farmers, local farmers, tourism entrepreneurs, environmentalists, and a local consumer group[23] member.

Planners Group

The key ideas developed by the planners group were the following.

- Recovery of knowledge about the value of local heritage. An ideal scenario would involve recovering natural and cultural values that Aranjuez once enjoyed: regenerate the rivers, the riverbank vegetation, the biodiversity, the market gardens, the historic roads and ways, the agricultural landscape, the irrigation systems etc.
- A quality trademark. Support the historic value of Aranjuez, based on agricultural quality in general—not just market gardens, but also grapes, olives, and other produce.
- Achieve a balance between production for local consumption and surplus for other markets. At present consumption is not local.
- Tourism. Take advantage of the fact that Aranjuez is already an important tourist destination, and link this tourism to its agriculture.

---

[23]Consumer groups are groups of citizens who organize themselves to buy products directly from suppliers on the basis of their shared values (typically ethical, organic, or local).

FIGURE 3.37

Poster describing three future scenarios for the peri-urban agrarian areas in Aranjuez in the year 2030.

To achieve this vision, the group made the following suggestions.

- Develop a strategy for regeneration of the rivers and restoration of riverbanks, and solve the issues with public-domain lands (debate about regulation at the source, with water diversion[24]).
- Strive to attain a productive agricultural landscape based on quality and consolidated in the collective imagination, as with, for example, the Jerte Valley.[25]
- Promote good agricultural practices and local consumption, but also take advantage of the market in the city of Madrid.
- Plan for valorization of production, and strive to obtain economic viability. Not everyone considers this indispensable—there are mechanisms for obtaining payment for services.
- Create public drivers of change: bonus payments for certain crops, and offer of public land to new farmers with a contract that includes conditions (environmental practices and production quality).

Overall summary: the group regarded this scenario as "ideal," but also achievable.

Civil Society and Farmers Group

This group oriented their scenario discussions around ES, eventually proposing two scenarios: diverse local supply and continuity.

The diverse local supply scenario focused on diversifying and increasing the quality of food production in particular. Specific aspects of this scenario included the following.

- Development of an agroecology perspective with a stronger emphasis on health and well-being through a culture of respectful enjoyment of the environment.
- Promotion of short supply chains, food based on proximity, and recovery of horticultural areas.
- Recovery of genetic heritage, traditional seed varieties, fruit cultivars, etc.
- Maintenance of agrarian infrastructure (irrigation canals, warehouses, etc.) and communication infrastructure (paths and droveways) to enable farming activity to continue.
- Correction of nitrate levels in the water.

Against the supply scenario described above, participants also developed a scenario emphasizing continuity of existing tendencies. In this scenario the future is in the hands of current market policies (related to the recent CAP reform), under which

---

[24]The Tajo—Segura canal, one of the largest hydrological infrastructures in Spain, transfers water from the River Tajo south to the River Segura, facilitating agricultural production in southeast Spain. The Tajo—Segura transfers continue to be a source of fierce debate on grounds of environmental impact.
[25]The valley of the River Jerte, in the province of Caceres, Extremadura, is well known for its agricultural production, in particular cherries.

there is limited scope for crop diversification in the area since land parcels larger than 10 ha already have their cropping pattern established. However, the group agreed that opportunities exist for the kind of developments proposed in the diverse local supply scenario in parcels smaller than 10 ha in size.

3.2.4 CASE 3.2.4: AGRICULTURAL LAND AND FAMILY FARMING: ENVIRONMENTAL PERCEPTION OF FARMERS IN LAS VEGAS COUNTY[26] (2012–2013)

3.2.4.1 Project Synthesis

In this project we carried out a comprehensive assessment of family farming and agriculture in the Madrid region to get to know the families' views about the environment and the future prospects for agriculture. The project is part of a program to develop sustainable agriculture in the Madrid region by training and capacity building for the agricultural population in innovative farming practices with potential to reduce impacts on the environment. One outcome of this project was the formation of Agrolab, a critical forum for directing agriculture in Madrid toward more sustainable pathways (see Agrolab Madrid for further information, https://agrolabmadrid.com). Work was conducted in two specific subzones of Las Vegas County, the Las Vegas area and the Alcarria area.

The main objectives of the project are listed below.

- To characterize the farmers in Las Vegas County and collect information about their perceptions of agriculture, their key motivations for choosing this activity, the role of farmland in general, and their view of the environment and environmental impacts of agriculture.
- To improve farmers' understanding of the environmental impacts of agriculture and encourage them to look for more sustainable practices (Fig. 3.38).
- To propose, design, and implement a collaborative action plan with local famers that contributes to solve environmental problems related to farming in Las Vegas County.
- To draw up a proposal for good practices, based on the strengths and challenges for the county's farmland, for training and capacity building by the regional government in support of family farming.

To achieve these objectives we carried out a qualitative analysis in the municipalities of the county, using a wide variety of techniques, to identify the most relevant problems and look for possible solutions along with local territorial stakeholders. Finally, recommendations for public strategic actions at the regional level were made to address the problems identified.

---

[26]Carried out by OCT under the leadership of IMIDRA, financed by IMIDRA.

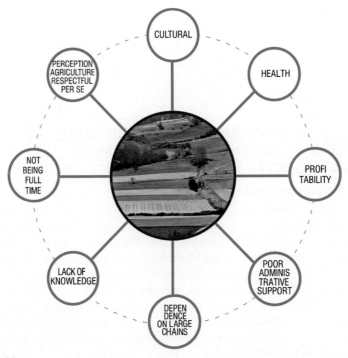

FIGURE 3.38

Environmental friendly practices in agriculture: motivational diagram.

3.2.4.2 Description of Techniques
3.2.4.2.1 Stakeholder Mapping (Chapter 2, Section 2.1.2.1.1)
The initial phase of the project comprised identifying stakeholders and key informants, and constructing a database to help planning of future work (Fig. 3.39).

3.2.4.2.2 Surveys and Interviews (Chapter 2, Section 2.1.2.1.2)
Semi-structured interviews were conducted, with the following objectives.

- To identify relationships between the different stakeholders (in particular between farmers and IMIDRA) and the way that the stakeholders perceive these relationships.
- To understand the farmers' perception of agriculture, the environment, and generational renewal, as well as the motivations, barriers, and incentives that led them to start farming.
- To raise awareness among stakeholders about the severe environmental impacts of farming in the area.

Personal Details (only available for work team)

CODE	Role
INS1	Agrarian Office South Madrid
INS2	Town council Aranjuez
INS3	Agrarian Office SouthEast Madrid
AE-cert1	Organic Farmer Vega
INS4	Research Institute
AC1	Conventional Farmer Vega
INS sin asoc1	Town council Villaconejos
INS sin asoc2	Farmer Trade Union, agrarian and livestock organizations coordinator (COAG)
AE-cert2	Organic Farmer Vega
AC2	Conventional Farmer Alcarria
AE-cert3	Organic Farmer Alcarria
AE-cert4	Organic Farmer Vega
INS5	Town council Perales de Tajuña
AC3	Conventional Farmer Vega
INS6	Town council Fuentidueña del Tajo
AC4	Conventional Farmer Alcarria
ASOC5	Irrigation Association
INS7	Organic Farming Committee
INS8	Wines of Madrid Committee
SIN ASOC3	Farmer Trade Union young farmers' association (ASAJA)
SIN ASOC4	Local Action Group for Las Vegas County and Alcarria (ARACOVE)

FIGURE 3.39

The database of stakeholders and key informants.

The interview outline used is shown in Table 3.3. This outline was adapted according to the profile of the person interviewed (public servants, conventional farmers, organic farmers, citizens, etc.).

Note that the interview is oriented around the perception of the farmer as a contributor to environmental degradation. Though this view is strongly supported by available scientific evidence, interviews of this type need to be conducted with sensitivity to avoid apportioning blame, which may lead to misunderstandings and distract from the stakeholders' key role as valued collaborators in identifying and solving environmental problems.

3.2.4.2.3 Discussion Groups (Chapter 2, Section 2.3)
A discussion group is a qualitative research technique that consists in grouping together 6–10 strangers with the aim of sharing and discussing their personal experiences and thoughts about particular issues. The technique is less direct than an interview and allows more room for contextual discussion. Four discussion groups were formed, for participants younger than 40 years and older than 40 years in each of the two study areas. Table 3.4 shows the verbatim transcript taken from one of these discussion groups.

3.2.4.2.4 Triangular Groups (Chapter 2, Section 2.4)
Two triangular group discussions were held, one with women farmers and the other with organic farmers. The triangular groups aimed to explore the in-depth

Table 3.3 General Interview Guideline

Interview Element	Objective of Element
Part 1: Introduction to the project	To introduce the project to the participant and explain its objectives and underlying motivations
Part 2: Basic data (name, age, gender, municipality of residence, main occupation, membership of associations, and level of formal education)	To obtain, in confidence and with full consent of the interviewee, the basic personal data necessary for analysis
Part 3: Regarding environmental awareness and concerns, elicit responses to the following. **1.** What is the environmental situation like in the area? Focus attention on main problems detected **2.** What are the causes of this situation? Focus attention on farming, and whether this is acknowledged as one of the major causes of environmental problems (Identify both extrinsic (not depending on the farmer) and intrinsic factors **3.** What are the consequences of these problems? Focus on environmental impacts of farm production	To understand how farmers view the environmental impact of farming, assess the degree to which they take responsibility, inform them of the severe negative environmental impacts of farming in this area, and communicate the extent to which this affects the famer directly
Part 4: Farming practices and farm management Questions about the different types of farm and farm management in the area, keeping in mind the key differences between crops in the two localities (Las Vegas and Alcarria)	To understand the attitude and long-term vision about farming practices in general

environmental impacts and the role of women and organic farmers in addressing them (Table 3.5, Fig. 3.40).

3.2.4.2.5 Discourse Analysis (Chapter 2, Section 2.1.2.2.6)

Participants' discourse was analyzed by matching transcribed verbatim dialogue arising from interview questions or group discussions with the key indicators for understanding the level of knowledge and awareness about the environmental impacts of agriculture. For example, stakeholder discourse about environmental damage to olive groves resulting from incorrect use of sewage treatment plant waste as fertilizer was matched to the indicator "perception about environmental problems in the county" (see Table 3.6). Thus although the selection of material from the (very extensive) verbatim transcripts did contain a degree of subjectivity, the primary information (stakeholders' literal discourse) on which the results of the project are

Table 3.4 Verbatim Transcription of Discussion Group Dialogue

---

Minute 1:50 to 4:20.

Discussion about participants' perception of the environmental situation in the area. There are two facilitators and six farmers in this group. This discussion fragment deals with the main problems related to natural resources.

Facilitator 2. All right, great. Look, the first thing we wanted to ask you is if you have problems in your area or municipality regarding natural resources' use in agriculture. That is, regarding natural resources' quality, or problems, in your municipalities. Resources like water, soil.

Farmer 3: Water

Facilitator 1: In what sense?

Farmer 6: Our [farm] is located in Guadalajara [just across the county border from the Alcarria area], so if it rains during the rainy season, and reservoirs are full, then we can grow water-intensive crops like maize. But if it doesn't rain, we have to leave the land in fallow, or grow another crop that doesn't give enough profit.

Facilitator 1: Any other issue?

Farmer 2: There aren't any other natural resources...

Farmer 1: For me, thinking about the vegetable plot, water quality. Now it comes down with soaps in it, and so the peppers, for instance, go down the drain.

Facilitator 1: And any other issue about natural resources useful for agriculture? Soil, fauna, I don't know, the quality...

Farmer 6: Fauna? There are a lot of rabbits.

Farmer 3: Lots of rabbits and wild boars, they ruin the maize, all kinds of crops. They devastate everything.

Farmer 4: Rabbits... For instance in Chinchón, you tell someone in Chinchón that there are rabbits in the Vega then the hunters come and take the land off you for hunting. And in Ciempozuelos they're there all the time, it's over the top.

Farmer 1: In the vegetable plot we respect them in general, but people eat them from time to time.

Table 3.5 Verbatim Transcription of Women Farmers' Triangular Group Dialogue

---

Facilitator 1: OK, I'm going to connect the video camera. Anyway, if you like, we can introduce ourselves, OK, so, your names, and then what kind of farming you do, and the crops you grow, just that. Starting over here, for instance. Just that, your name and the kind of crops you have, just to get to know each other a little.

Farmer 1: C.H. I'm mainly dedicated to grapevines. I have a vineyard and a winery, they're organic. And also an organic olive grove. And cereal crops, but they're conventional.

Farmer 2: And why? Why is it not organic?

Farmer 1: Well, actually it's not organic because I haven't registered it with the committee; I don't know how well it would sell and if it's worthwhile... I mean if it's worth paying the registration fee. I haven't studied the marketing possibilities of grain. I would have to get in touch with a livestock farmer.

Farmer 3: So it doesn't have the label. But the crop itself is organic or is treated?

Farmer 1: It's conventional. I don't buy an organic fertilizer, as I do in the vineyard. Then, what happens... it would be good [to do]... But, but I know, I think that in Madrid I only know one livestock producer that's organic, with organic cattle. I don't remember the name...

FIGURE 3.40

Working session with participants in one of the triangular groups.

based appears with the results in the same table. Clearly showing the primary evidence and the way this links to the results helps to maintain transparency and guard against conscious or unconscious bias on the part of the researcher.

3.2.5 CASE 3.2.5: INTEGRATIVE SYSTEMS AND BOUNDARY PROBLEMS[27] (2006–2009)

3.2.5.1 Project Synthesis

As seen in the preceding case studies, actors like farmers, urban developers, regional and local government planners, and citizens' groups frequently come into contact with each other over the best use of land and resources. The different beliefs, values, and world views held by such actors, known as *boundary judgments*, often translate into intractable conflicts over the way land and resources should be managed. The main focus of our work under ISBP was to explore the relationship between these beliefs and the sociopolitical and geographical context in which each actor operates through case studies of the peri-urban areas of two large cities, Madrid and Barcelona. The outstanding issue at that time was the very intensive urban development that was taking place and the apparent lack of appropriate land management models for these areas (Cartledge, Dürwächter, Hernandez-Jimenez, & Winder, 2009;

[27]Funded by the Sixth Framework Programme of the European Union (FP6). Project coordinated by Newcastle University (United Kingdom). This work was undertaken with the Department of Rural Planning and Projects, College of Agricultural Engineering, and coordinated by David Pereira Jerez, Polytechnic University of Madrid (Spain).

Table 3.6 Process of Collection and Analysis of Stakeholder Discourse

Study Variable	Indicators	Associated Question in Interview or Discussion Group	Key Issues	Interview with Local Politician		Discussion Group With Organic Farmers	
				Verbatim Text	Key Ideas	Verbatim Text	Key Ideas
Environmental awareness	Perception of environmental situation in comarca Las Vegas	What is the environmental situation in the comarca Las Vegas?	External inputs	"They are using too much herbicide in Aranjuez. In Chinchón, they are applying sewage as fertilizer and this is very devastating for olive trees"	Natural resources: water. Conflicts about water between villages and towns. Excess of chemical inputs in crops and soil	"The environmental situation has improved a lot, although it depends on the area of the region. For instance, olive trees and vineyards have reduced the chemical treatments considerably and now it is quite easy to manage an ecological farm on olive and vineyards here"	Olives and vineyards are easy to manage under organic farming practices
	Perception of environmental causes in comarca Las Vegas	What are the main causes of this situation? Are they linked with agrarian practices? Are there other external factors?	Polluted agrarian production associated with environmental problems	"Traditional agrarian practices here follow rotation of crops which avoid bad treatment of the soil. The soil recuperates its nutrients with the annual change between different crops"	Crop rotation key for a healthy soil: cereals (corn, wheat, barley), pastures and forage, legumes and root vegetables	"Nowadays, Colmenar and Villaconejos are almost 100% organic farming, although there is a wide range of agrarian practices depending on the crops. An important issue is that organic famers are not seen as mad people anymore"	Agrarian practices are directly related to environmental situation in water and soil
Training needs	Knowledge transfer and learning process in agrarian practices	How did you learn the agrarian activity? Family tradition? Training courses?	Knowledge transfer and learning process in agrarian practices		Family tradition supported by training in new varieties, environmental practices, and special support in marketing products	"We are self-trained in organic farming"	Official training is non-existent in organic farming

Hernandez-Jimenez & Winder, 2009; Hewitt & Hernandez-Jimenez, 2010). Work was strongly oriented toward dissemination of knowledge to raise the level of awareness of these issues in society. The economic crisis of 2007, which began to be strongly felt in Madrid around a year later with the collapse of the construction industry, offered an opportunity to reflect on the unsustainable speculative development of previous years and look for new integrative approaches to the planning of the territory.

In practical terms we looked to identify convergence mechanisms for more sustainable land planning and worked to provide spaces for reflection and rapprochement between stakeholders. The culmination of the work was a seminar cycle entitled "When the city grows: Seminars of analysis and reflection about the land planning model of the Community of Madrid." The seminars brought together civil society representatives, scientists, land planners, and politicians for discussion and debate about land management in the Madrid region. They were held in various locations in Madrid city, culminating in an event hosted at the Caixa Forum, a prestigious cultural venue near the Prado Museum. The visibility of the event and venue ensured a good level of representation from public officials and politicians from the regional assembly and national environment ministry.[28] These events, and the subsequent participation of the authors in national conferences and appearances in the media (http://observatorioculturayterritorio.org/wordpress/?page_id=123), provided the platform for our group, OCT, to become established as an independent not-for-profit organization (see http://observatorioculturayterritorio.org/wordpress/, a Spanish website with a downloadable information sheet in English).

3.2.5.2 Description of Techniques

3.2.5.2.1 Digital Visualization and Multimedia (Chapter 2, Section 2.1.2.3.4)

To disseminate the range of issues identified in the peri-urban spaces of our case study cities, an exhibition was held covering the following thematic areas.

1. When the city grows.
2. Imagining the future.
3. Stakeholders: decision-making complexity.

---

[28]"With the aim of linking science and policy as the only valid means of obtaining an efficient land planning model, the seminars will be attended by diverse specialists (in rural planning, environment, economy, geography etc, and representatives of the main political parties in the Madrid Assembly (PP, PSOE, IU), as well as personalities from the political—institutional sphere in Madrid and the National Environment Ministry." Madrid Polytechnic University press release, May 2009, available at http://www.buscagro.com/blog/1190-nota-de-prensa-jornadas-de-analisis-y-reflexion-sobre-el-modelo-de-gestion-territorial-en-la-comunidad-de-madrid/. This quote from the press release gives a flavor of the event, even if we are not in full agreement about the need for land planning to be more "efficient." The presence of opposing political parties, the Environment Ministry, environmentalist groups, and high-ranking public officials like the director of planning for the city of Madrid ensured many extremely lively debates.

4. Madrid region, the countryside at the urban fringe.
5. Peri-urban spaces, urban—rural transition.

   Figs. 3.41 and 3.42 show the display panels for two of these thematic areas.

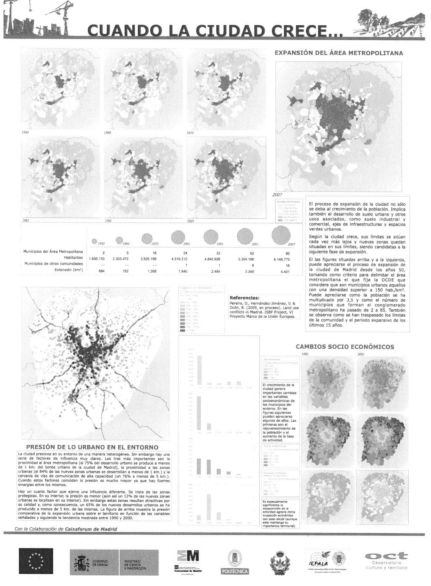

FIGURE 3.41

Display panel for the thematic area "When the city grows."

ESPACIOS PERIURBANOS
Transición de la ciudad al campo

Un viaje desde la ciudad al campo

Los espacios periurbanos actúan como puente entre los núcleos urbanos y rurales, cumpliendo una función de equilibrio territorial entre la ciudad y el campo. En las últimas décadas estos espacios han absorbido las condiciones de las áreas urbanas, relegando así su carácter de transición entre ambos mundos.

¿Son los espacios periurbanos la última oportunidad en las relaciones urbano-rurales?

Se analizan las oportunidades que los espacios periurbanos, como espacios intermedios y conectores del medio rural y urbano, tienen en el territorio y las funciones relevantes que pueden desarrollar en el mismo. También se explora la importancia de la agricultura y del medio rural en el entorno de las grandes y pequeñas ciudades por su carácter estratégico y funcional, bajo el contexto actual de crisis.

Problemáticas en las relaciones urbano-rurales

El crecimiento urbano no sólo implica la construcción de viviendas, sino también la expansión de las infraestructuras, principalmente la red de comunicaciones, el incremento de zonas comerciales e industriales, que fragmentan el territorio y estrangulan estos espacios de transición, que en la mayoría de las ocasiones suelen ser espacios agrarios.

Área urbana y sus flujos de comunicación

Entorno periurbano, amortiguando los efectos de la ciudad sobre el campo

Fragilidad de los espacios de agrarios en la relación campo-ciudad

Además de las circunstancias que influyen en los cambios de usos del suelo, el marco político ha podido favorecer la situación de vulnerabilidad de estas zonas. Las Leyes del Suelo, tanto nacional como regional, han generado unas dinámicas de liberalización del suelo, que han provocado un crecimiento de las zonas urbanizadas en la periferia de las ciudades, en detrimento de los espacios agrarios existentes. Por otra parte la Política Agraria Común ha generado dinámicas de abandono de la agricultura tradicional y familiar asociada a pequeñas superficies, dando paso a una agricultura mas industrializada de mayores extensiones. Las políticas de desarrollo rural, también han dado lugar a un tipo de turismo en el que los espacios rurales pasan a ser mero soporte físico del ocio y del consumo de los habitantes de la ciudad.

Todo esto ha llevado a la consecuente modificación del paisaje agrario, con una ambientalización del campo, generando así un sentimiento de desarraigo por la tierra y fatalismo hacia lo rural.

Con la Colaboración de Caixaforum de Madrid

¿Qué espacio periurbano queremos demandar a nuestros Gobiernos Locales?

¿Cuál es tú espacio periurbano ideal?
- un espacio agrario para cada área urbana -

¡CUÉNTANOSLO!

Principales funciones de Agricultura Periurbana

Equilibrio territorial
Caminos rurales en Belmonte de Tajo

Carácter productivo
Cultivando hortalizas en Zarzalejo

Educación agro-ambiental
Jornadas de Trashumad de la Comunidad de Madrid

Espacios verdes para uso recreativo y ocio

Vía pecuaria en el Valle del Lozoya

Huertas de

Valorización cultural
Cabañas de ganado y muros de piedra la Sierra

en Mad

Referencias:
V. Hernández-Jiménez y B. Ocón. Mesa Redonda "Espacios Naturales Periurbanos" en CONAMA9, Diciembre 2008, Madrid.
V. Hernández-Jiménez, B. Ocón y J. Vicente. Espacios Periurbanos, transición de la ciudad al campo. Ecosostenible, nº 49, Marzo 2009.

FIGURE 3.42

Display panel for the thematic area "Peri-urban spaces, urban—rural transition."

3.3 THEME 3.3: CONFLICTS, CITIZENS, AND SOCIETY: PARTICIPATORY MODELING FOR A RESILIENT FUTURE

INTRODUCTION

The Search for New Models

In the first part of this chapter, *Theme 3.1: Getting to know the territory*, we saw how a range of different participatory tools and approaches can be used to engage local communities in restoring a sense of value, self-esteem, and identity to their locality by bringing distinctive local landscapes, historic elements, products, or customs to the fore and learning how to appreciate them in our cynical and disconnected modern world. In the second part of the chapter, *Theme 3.2: Between city and country*, we looked at how the problem of increasing alienation of our towns and cities from the formerly productive landscapes that surround them, which has arisen as a result of the globalization of the food network, can be directly addressed by community-level interventions in agroecology and restoration of productive activity as a means to protect key ES. These approaches are particularly relevant on the peri-urban fringe of large cities like Madrid and Barcelona, where the peculiar economics of speculative development leave formerly valuable agricultural areas as virtual wastelands, awaiting their turn to be built on. Yet both these aspects of what we have called "a culture of the territory" need to be part of a broader agenda for change if they are to be successful beyond the level of the local community. The need for overarching new visions, new ways of looking at and valuing the territory that progress beyond the conventional "growth without limits" model of neoliberal capitalism, is clear, and that is the case we make in Chapter 1 of this book. But linking these two aspects, combining the local-scale, mostly practical, community-led initiatives with a new, nurturing (rather than extractive) model of the territory which seeks to preserve our common future and at present exists mostly only in theory, requires an enormous effort of imagination. Many people believe passionately in sustainable transport, but travel to and from work by car. Most of us are happy to believe that there are better alternatives to large-scale industrial farming, but find it impossible in practice to feed ourselves and our families exclusively from a family allotment or community market garden scheme. Cynics and defenders of the status quo always argue that this just shows that those of us who believe in a brighter, greener future for our planet are dreamers, sandal-wearing idealists who do not live in the real world. We should "get real," the argument seems to go, and, by logical extension, abandon any idea of a future that is not based on environmental destruction. Not only is this argument completely absurd, but it plays into the hands of the powerful, those who have the most to gain from the status quo, and generates a self-reinforcing perspective that nothing we do will ever change anything. Living within the bounds of what planet Earth can provide is not only perfectly compatible with functioning human societies, it is a necessity. Campaigners for the abolition of slavery were once regarded as idealistic and impractical, yet the practice is considered abhorrent today. If we are successful, future generations will no doubt view 21st century humanity's war on the planet with the same incomprehension that we look back on the Atlantic Slave Trade today.

Building Common Visions of the Future

To be able to move forward, then, we need to liberate ourselves from the idea that everything is hopeless, or, alternatively, that the current system is the best of all possible systems and there is no room for any improvement. The space to do this kind of thinking is certainly difficult to achieve as part of our daily lives, but it is not always easy within the constraints of a scientific project either. But opportunities can be found, as we hope to demonstrate through the three case studies we have selected for detailed description in the following pages.

In the final part of this chapter, *Theme 3.3: Conflicts, citizens, and society: participatory modeling for a resilient future*, we try to show how we can find this opportunity through a range of participatory knowledge-sharing approaches that are more ambitious, more extensive, and more tightly interconnected than in the case studies used so far. The goal is not to search for a single one-size-fits-all solution (surely it is that kind of thinking that got us here in the first place), but open our minds and our imaginations to a range of possibilities that might address the problem. If we have no idea what our future will look like, we have no way of finding the road. As Ignacio Palomo noted: "If we do not know where we want to go, it will be more difficult to arrive" (personal communication in a presentation given to Doñana stakeholders on participatory scenario planning). If we can build a common vision of the kind of future we want, land planning and decision-making might be oriented toward that end. This common vision may be different for everyone, but it is likely that we will agree on some important elements. Though somewhat overused in recent years by policy-makers, the idea of a roadmap remains relevant, although it is important to remember that many roads may lead to the same destination and one road may pass many destinations along the way.

Decision-Making for Our Future in the Territory

The development of roadmaps or guidelines to steer us toward the kind of future that we want is already a fundamental part of planning in a conventional sense. The main difference relates to whose information is used to define the path we take. It was long thought in some circles that a perfectly acceptable approach to planning a long-term strategy, say for transport, energy, or agriculture, was to gather together political representatives, big companies whose capital or technological expertise would be needed, and a number of expert advisors in, say, economics or agriculture. This model of decision-making, though it remains very common, is increasingly criticized. On the one hand it leads to a presupposition that planning decisions are one-off events, and once they are made they cannot be unmade. In fact, in many cases decision-making in the territory needs to be understood as a continuous process and not as a single action to be signed off by bureaucrats and implemented by unthinking workers. At the same time questions can be raised about the objectivity and transparency of this kind of decision-making by expert committee approach. Frequently land policy decisions are bound to ideological constraints, and not uncommonly relate more to the interests of large corporations or major landowners than to the interests of the citizen.

The main problem with this approach, apart from being fundamentally undemocratic, is that it dramatically restricts the range of options that are considered in the decision-making system. Thus the symptom of a system that is overreliant on planning by diktat, on the basis of shady bargains made in smoke-filled rooms, is paralysis. The system's insiders, or *incumbents*, have, or believe they have, so much to lose from any kind of change that they stubbornly resist all but the most gentle of measures. Thus in our Madrid case study, touched upon in the ISBP project discussed in the previous section, stakeholders had become locked into a model of speculative urban development (extraordinarily profitable for insiders) that prevented any kind of more holistic and consensual approach to the territory, and ultimately created a bubble that destroyed the Spanish economy (Naredo, 2008; Romero et al., 2012).

Our second case study, Doñana, is once again in crisis, ravaged by forest fires (June 2017) that caused the evacuation of el Acebuche, the visitors' center and park administration offices where we held our workshops some years before (El Diario, 2017). Though suspicion has mounted that the fire was deliberately started, the real issue at the heart of the problem is the refusal of successive regional governments to admit the fundamentally unsustainable trajectory of this natural protected area or contemplate any real alternative to the existing pattern of development in the hinterland of the park. The protected heart of Doñana, the marshland habitat, is dependent on water from the Guadiamar River, site of a disastrous industrial accident in 1998 following the collapse of a tailings dam at the mine of Aznalcóllar. In every direction the zone beyond the protected area is used for intensive crops like strawberries, citrus, cotton, and maize. The aquifer on which the habitat depends is massively overexploited, and many of the wells are illegal. One is left with the impression that while no one can countenance the idea that Doñana may be lost, its life-support systems are locked into an unsustainable pathway because of a fundamental lack of imagination on the part of policy-makers over many generations, and because, as we observed above, incumbents are prepared to do anything rather than risk a change of direction.

In our last case study, under the Complex project, we looked more broadly at the climate crisis and the European response to it with the Low-Carbon Roadmap (LCE). In particular we were interested in the way that crisis-hit governments like the Spanish administration of Mariano Rajoy have backpedalled from their commitments to low-carbon energy systems, again in great part because incumbents felt threatened by the loss of their profitable business model, not because any particular aspect of the clean energy transition is otherwise unworkable. Could this happen elsewhere, we wondered, and what must we as citizens do to get this process back on track?

The common thread running through all these cases, and the key motivation for the participatory approach we have followed in each case, is the feeling of crisis or deadlock, with the system in paralysis, unable to shake off the old model that has been so manifestly unsuccessful at protecting the life-support systems on which our planet depends, but also unable to take a new path. Frequently there are many more stakeholders involved in the problem in question than are formally acknowledged, and these unacknowledged or excluded stakeholders are a rich source of instability and conflict (Hewitt, Winder, Hernández Jiménez, Alonso Martínez, & Román Bermejo, 2017).

More than anything else, what we have tried to do, with varying degrees of success in the cases discussed here, is open up alternative pathways and push for greater inclusivity in those considered to be "stakeholders" in territorial planning. To this end, all the case studies selected for inclusion under this theme address in one way or another the co-development of models and tools to guide policies involving representatives of the stakeholder groups which most need them. Due to the high degree of complexity this kind of participatory work entails, all three of these cases come from research projects of 3 or 4 years' duration, TiGrESS, DUSPANAC, and COMPLEX, which provided a long enough timescale and sufficient resources to move from simpler, discussion-based stakeholder processes to more developed participatory modeling activities. The cases presented in this section are the following.

> Case 3.3.1: TiGrESS: Time—geographical approaches to emergence and sustainable societies. Funded by the European Union (EU) Fifth Framework Programme. Newcastle University, United Kingdom (coordinator), Universidad Europea de Madrid, Spain (work described here) (2002—2006)
> Case 3.3.2: DUSPANAC: modeling land-use dynamics in the Spanish network of national parks and their hinterlands. Funded by the Autonomous Body for National Parks (OAPN). University of Alcalá, Madrid (coordinator), and Observatorio para una Cultura del Territorio, Spain (2011—2013).
> Case 3.3.3: COMPLEX: Knowledge-Based Climate Mitigation Systems for a Low-Carbon Economy. Funded by the EU Seventh Framework Programme. Newcastle University, United Kingdom (coordinator), and Observatorio para una Cultura del Territorio, Spain (work described here) (2012—2016)

GENERAL APPROACH AND STRUCTURE

In this section, as in previous sections, we present the principal tools and approaches that appear in these three case studies. Fig. 3.44 shows the tools which are most relevant to and most frequently applied in the cases presented, all of which are described in detail in Chapter 2. In these projects, which are broader in scope than most of the previously discussed cases, we have followed a more organic approach which we tried to keep as flexible and open as possible. Broadly, the structure of our approach was as follows. We began work on each case with a knowledge-sharing and elicitation process designed to understand the history and evolution of the case at hand, whether it is Madrid's unsustainable development model, land-use change as a driver of land degradation in Doñana, or the trajectory of renewable energy development in Navarre. At this stage a wide range of basic techniques were used in combination: *surveys and interviews* (Chapter 2, Section 2.1.2.1.2), *flow diagrams* (Chapter 2,

Section 2.1.2.2.5), *geographical analysis of the territory* (Chapter 2, Section 2.1.2.2.7), *discussion groups* (Chapter 2, Section 2.1.2.2.3), and the *timeline technique* (Chapter 2, Section 2.1.2.2.1). Building on this collective learning process, we typically travel through a series of stages looking at the role of individual actors in the system, trying to understand their capacity for change, linking problems to possible solutions, and trying to unpick the chain of responsibility (e.g., *sociograms* (Fig. 3.43); Chapter 2, Section 2.1.2.2.2). In all three cases we have used a spatial modeling approach combined with a series of future narratives of change, called scenarios, which either respond directly to stakeholders' concerns arising out of the participatory process or are co-developed with stakeholders through detailed workshop activities. These activities logically involve *participatory modelling* (Chapter 2, Section 2.1.2.3.1), *scenario development* (Chapter 2, Section 2.1.2.3.3), and *participatory mapping*; Chapter 2, Section 2.1.2.3.2.2). Participatory processes of the kind of complexity and depth we see here often struggle to provide participants with the tools to evaluate objectively the process itself. However, this is key if we are to claim that the change of understanding implicit in successful social learning has really been achieved. In the DUSPANAC project we dedicated one of our three workshops to this kind of process evaluation, using the *dartboard technique* (Chapter 2, Section 2.1.2.4.3) and a *participation ladder* (Chapter 2, Section 2.1.2.4.2). Finally, the *SWOT matrix* (Chapter 2, Section 2.1.2.1.7) was used by the research team to evaluate the project after its completion, assess weak points, learn lessons, and look for future pathways arising from the work undertaken. Fig. 3.44 shows all the techniques used in this theme, and Fig. 3.45 shows how each technique relates to a specific project.

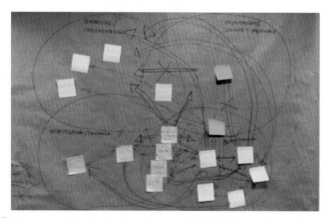

FIGURE 3.43

Participatory sociogram of actors and relationships. The *circles* show actors grouped by sector, while the *arrows* indicate relationships between actors. Post-it notes are participants' observations about particular actors or relationships.

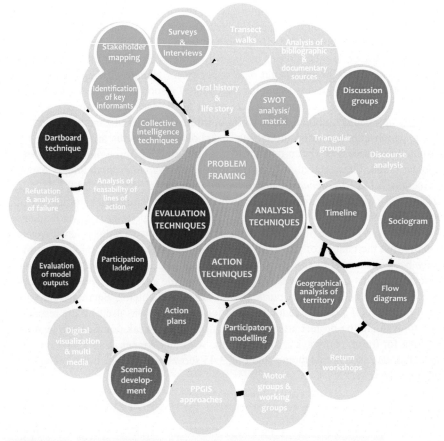

FIGURE 3.44

General methodological framework for Theme 3.3: Conflicts, citizens and society. The techniques most frequently used in this theme are highlighted.

DESCRIPTION OF PROJECTS AND TECHNIQUES USED

As for the previous themes, a synthetic review of each project included in *Theme 3.3: Conflicts, citizens, and society* is presented in the following pages, with a brief description of the techniques employed in each case. This review is presented as a factsheet containing the basic information about the project: its geographical location, context, funding body, main objectives, and a synthetic description of the work carried out. As in the previous themes, we remind the reader that we do not provide an exhaustive description of each and every technique used in a given project, but we have selected what we regard as the most important or illustrative techniques relevant to the theme for detailed treatment.

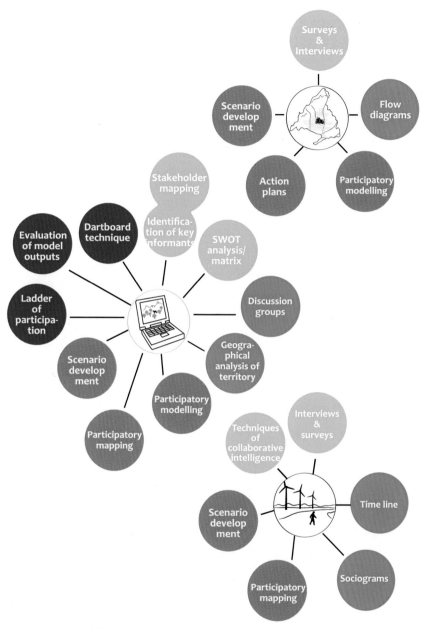

FIGURE 3.45

Diagram showing the relationship of the techniques used, following the color coding in Fig. 2.2 (Chapter 2), to the projects described in this theme. Each project is given a symbol, which appears in the top right-hand corner of each factsheet to help the reader identify the project.

3.3.1 CASE 3.3.1: TIGRESS: TIME—GEOGRAPHICAL APPROACHES TO EMERGENCE AND SUSTAINABLE SOCIETIES[29] (2002—2006)

3.3.1.1 Project Synthesis

Tigress was an international interdisciplinary EU-funded research project carried out between 2002 and 2006 which aimed to facilitate the shift toward more sustainable development models in Europe by bringing together disparate stakeholder communities to try to resolve conflicts (Winder, 2006). In the Madrid region, an initial appraisal of the effectiveness of the land planning system in securing sustainable development revealed serious deficiencies, in particular, a generalized lack of compliance with European directives and guidelines on the environment and a high level of conflict between groups of antagonized stakeholders (Fig. 3.46). The principal failing identified by nearly all stakeholders consulted was a regional land planning model based on speculative urban development which failed to take into account any other criteria. The resulting disaffection, mistrust, and

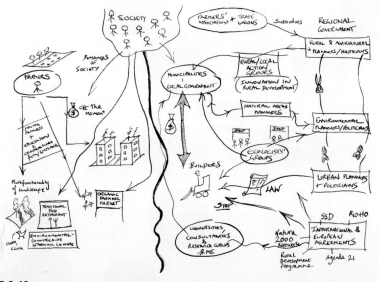

FIGURE 3.46

Researcher's rapid "rich picture" appraisal of the land planning situation in Madrid at the time the work was carried out (2005).

[29]Fifth Framework Programme of the European Union (FP5) (project coordinator: Nick Winder), Polytechnic College of the Universidad Europea de Madrid (work package leader Mª Soledad Garrido Valero), and Social Sciences Faculty, Newcastle University (United Kingdom).

disagreement between civil society and other actors like planners and private sector developers about the regional development model meant that progress on sustainable development had stalled. The key aim of the Madrid case study was to search for opportunities to make progress by engaging with a range of key stakeholders to diagnose problems and propose policy interventions.

In particular, the team was interested to investigate the role that sustainable agriculture might play in the protection of key ecosystems and the recovery of ecological and landscape value in areas under intense pressure from urban development. A participatory process was initiated in three case study areas of Madrid, the municipalities of Belmonte de Tajo, Cercedilla, and San Sebastian de los Reyes (Hernández Jiménez, 2006). These municipalities had both intense pressure from urban development and a strong and active local organic farming community. The main techniques used in the different stages of the project are described in the following subsection.

3.3.1.2 Description of Techniques
3.3.1.2.1 Surveys and Interviews (Chapter 2, Section 2.1.2.1.2)
Our first steps in this project were to establish contacts with the stakeholder community, locate their farms, and investigate their system of interests and the relationships among the stakeholders involved in local development. Stakeholders included individuals and groups involved in environmental management and land planning, and influencing the approaches to decision-making; they included farmers, associations, landowners, rural development groups, land planners, and local government officials. We began with a reconnaissance phase of research directed toward two stakeholder communities, each operating at a different scale. These were municipal (local) and regional policy-makers (Table 3.7).

Contact was made with these stakeholders through personal visits during which semi-structured tape-recorded interviews were carried out. Interviews comprised two main types, depending on the scale of action of the interviewee (local or regional). For local-scale stakeholders, the focus was on understanding their perspectives on land use and organic farming. For the regional policy-makers, interviews were mainly oriented toward land planning strategies, tools, and instruments. Interviews took place in a wide variety of locations: farms, cooperatives, bodegas, town councils, horticultural nurseries, and headquarters of the regional government. In all, 22 semi-structured interviews were conducted, and after initial analysis of interview results follow-up visits were made to all interviewees for informal, more in-depth, conversations on topics of interest.

For the local-scale stakeholders (45% of the total interviews), the topics most frequently mentioned (and hence most significant for participation) concerned the situation of farming (46%), political issues (31%), new development initiatives (31%), and participation (31%).

For the regional-scale interviews (55% of the total number), the most frequently mentioned topics were landscape multifunctionality in regional development (44%), knowledge of the administrative and decision-making system (33%), and knowledge of and/or participation in land planning regulations (55%).

Table 3.7 TiGrESS Project Stakeholders

Relevant Stakeholders in Local Community	Relevant Stakeholders in Regional Community
3 organic farmers in Cercedilla, Belmonte de Tajo, and San Sebastián de los Reyes 2 conventional farmers in Belmonte de Tajo and San Sebastián de los Reyes 1 cooperative livestock farm in Cercedilla 1 forestry agent in Cercedilla 2 SMEs: bodega (winery) and nursery garden 1 leader of the Organic Farming Association (APRECO) 2 mayors: Cercedilla and Belmonte de Tajo	2 technicians and planners in Departments of Rural Development and Environment within the Regional Directorate 1 natural protected area manager of a site of community interest, Curso Medio del Rio Jarama, in San Sebastián de los Reyes 1 leader of Regional Committee of Organic Farming 2 farmers trade unions: COAG and Plataforma Rural 4 local and rural development groups:[30] The intermunicipal community group for southeast Madrid (MISECAM) and Cercedilla Town Council (ADLs) and ARACOVE and Sierra de Guadarrama development association (ADESGAM) (GALs)

3.3.1.2.2 Flow Diagrams (Chapter 2, Section 2.1.2.2.5)

Following the interview phase of the work, a problem identification process was initiated using a method known as the logical framework approach (LFA), commonly employed in preparation, implementation, monitoring, and evaluation of rural development projects (Gasper, 2000). The core of the LFA involves the construction of two types of flow diagrams, the *problem tree* and the *objective tree*.

Problem Tree

The information obtained from stakeholders in the interviews was used to sort problems into a hierarchy. A classification of the main fields of interest suggested the themes of *institutions and committees, participation, scientific information, and planning and management tools*. The main problems identified in the first level of the hierarchy were as follows.

- *Institutions and committees:* poor vertical communication between local (socalled municipal) and regional institutions was a significant factor.
- *Participation:* inappropriate identification of stakeholders and the short time intervals during which they can participate in the process are considered the major problems. Many opportunities to engage stakeholders are also missed.
- *Scientific information and accessibility:* the absence of reliable scientific information that can provide an overall territorial perspective is perceived as a problem.

---

[30]These are known both as local action groups (LAGs) in English and as GALs and agentes de desarrollo local (ADL) in Spanish.

- *Planning and management tools:* there are too many tools for territorial planning and management, and they have not kept abreast of international developments. Consequently, convergence with national and international legislation is slow and compliance is poor.

The definition of problems in this way led us naturally into the exploration of an *objective tree* and the options for resolving them.

Objective Tree

We recall from Chapter 2 that the objective tree is the opposite of the problem tree. When we define a problem tree we are looking for obstacles; when we move to considering objectives, we look for possible solutions.

- Institutions and committees: the institutions are working independently on the territory, and thus their actions over the territory lack coherence and strategic monitoring. The objective tree developed is shown in Fig. 3.47.
- Participation: the unsuitable participation process is caused by inadequate stakeholder identification and the sparse level of participation. The objective tree is shown in Fig. 3.48.
- Scientific information and accessibility: the non-existent reliable scientific information from the social, environmental, and economic points of view led to the development of the objective tree shown in Fig. 3.49.
- Land planning and management tools: inefficiency in the implementation of land planning tools as well as incomplete compliance with international directives and guidelines led to the development of the objective tree shown in Fig. 3.50.

In this way interviews and follow-up conversations with key stakeholders at the local and regional levels allowed the research team to identify the key problems in

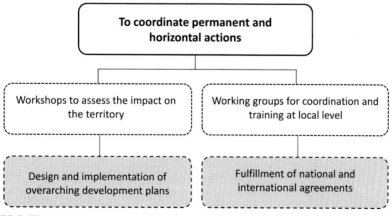

FIGURE 3.47

Objective tree for solution of problems related to institutions and committees. (Specific actions in gray.)

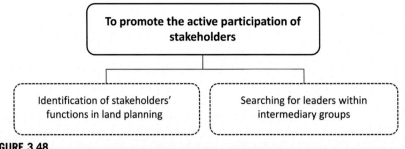

FIGURE 3.48

Objective tree for solution of problems related to participation.

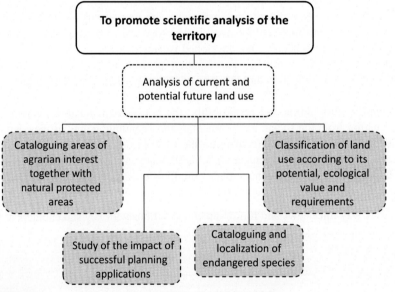

FIGURE 3.49

Objective tree for solution of problems related to scientific information and accessibility. (Specific actions in gray.)

FIGURE 3.50

Objective tree for solution of problems related to land planning and management tools.

land planning in Madrid, and link these problems to objectives that it was hoped would lead to their solution. In the final stage of this exercise the research team proposed seven policy interventions arising from these results to help move land planning in Madrid on to a more sustainable footing (Fig. 3.51).

To find out how stakeholders might respond to these proposals, a further set of interviews solicited their views in response to the following questions.

- What expectations do you have for each of the seven initiatives? (Would you veto?)
- Opinions about the proposed initiatives (in favor/against).
- What benefits could you obtain from them (direct benefits or externalities)?
- What kind of resources could you devote to the initiative (at what cost)?
- What conflict of interests could the initiative create (negative externalities)?
- What sorts of links do you recognize between initiatives?
- Given the list of possible initiatives, have we missed any important stakeholders?

Each session required an intensive explanation of the policy interventions and multilevel decision-makers involved in sustainable land planning. The trust we had already built up with stakeholders was particularly valuable here, because they were willing to listen patiently and ask probing questions. These were generally friendly but very business-like meetings. Unambiguous questions were asked, like "would you support or veto this action?", and consequently very clear opinions were expressed. Discussion frequently focused on stakeholder conflicts, which tended to break the flows of information between the key actors in the multilayered decision-making system. This was the stage in our fieldwork where the stakeholders participated most effectively. Each took positions on the policy interventions proposed and contributed strongly to the development of an effective planning network.

All the stakeholders consulted believed that construction companies were the largest landowners in Madrid. However, the absence of these actors from this work was striking—they were invited to participate, but chose not to do so. These

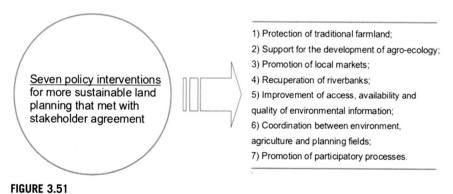

FIGURE 3.51

Seven policy interventions for more sustainable land planning in the Madrid region. Hypothetical consequences of these policy interventions were simulated in a land-use model of the Madrid region which was developed in response to stakeholders' concerns. Adapted from Hernández Jiménez (2006).

companies also have control over the municipal land planning schemes[31] in Madrid, especially in the lower-budget municipalities. Developers commonly lobby mayors to win concessions on future planning schemes before the planning zones are fixed, so the level of regulation is often minimal and key decisions may be made before open consultation is invited.

3.3.1.2.3 Participatory Modeling (Chapter 2, Section 2.1.2.3.1) and Scenario Development (Chapter 2, Section 2.1.2.3.3)

The participatory work described in the preceding subsections generated four critical problems in the problem tree of the LFA.

1. Institutions and committees.
2. Participation.
3. Scientific information and accessibility.
4. Land planning and management tools.

The stakeholder engagement process addressed the first two of these problems, generated seven policy options, and negotiated substantial "buy-in" from key stakeholders at five levels in the governance hierarchy (municipal, intermediary, regional, national, and supranational). This unexpectedly positive result in a situation which we had reasons to believe could easily have become confrontational gave us the impetus to tackle the last two challenges identified by our participant network.

To achieve this aim, a land-use model[32] known as the Madrid policy simulator was developed for the whole Madrid region to visualize the possible future consequences in the territory of the policy interventions that emerged in the previous phases of work. Although we ultimately hoped that the model would be of interest more widely outside the Madrid region, our key end users were the stakeholders we had already engaged in previous phases of work. Modeling work was completed in 2006.

The model represents land use in the Community of Madrid (CM) in the form of a standard two-dimensional cell-based (raster) map familiar to users of geographic information systems (GIS). Land-use base mapping was produced by researchers in Madrid from remotely sensed data for three years (1989, 1997, and 2002). From a starting configuration derived from a real land-use map (in this example, land use in the CM in 2002), the model advances through a series of stages representing time steps. At each step land use across the entire map is recalculated on the basis of transition rules determined for each individual cell, which takes the land use for which its transition potential (TP) is highest. The TP for each land-use cell in the model was calculated on the basis of the interaction of four key model parameters: *neighborhood dynamics* (cell neighborhood rules determined from previous

---

[31]Plan General de Ordenación Urbanística (PGOU) in Spanish.
[32]The land-use model was developed using the well-known Geonamica modelling framework (now usually known as Metronamica), developed by the Research Institute for Knowledge Systems in Maastricht, Netherlands.

empirical study of land-use dynamics, e.g., repulsion of industry on housing, attractiveness of farm land for urban use, etc.), *accessibility* (distance from infrastructure like roads and railways), *suitability* (physical or geographical determinants, e.g., mountain areas are often more suited for pasture than arable farming), and *zoning* (areas protected or reserved for certain land-use types). A random variable (α) was included to account for the effect of unknown and chance factors in land-use allocation, such that no two simulations will ever be identical even where they have the same transition rules. For detailed description of the functioning of the Metronamica model please see www.riks.nl/resources/documentation/Metronamica%20documentation.pdf.

The model's effectiveness for representing real land-use dynamics was tested by a process known as calibration, in which land-use simulations are generated for a historic date and compared with real land-use maps for the same date. A range of alternative options to the current regional planning model were considered, in the form of five future scenarios of land use in the region in 2025 (Hernández-Jiménez, 2007; Hernández-Jiménez & Winder, 2006). A baseline scenario (Scenario 0, business as usual) was also developed for 2025, to represent the effect of a *laissez-faire* approach to land planning in the territory (no change to current planning policy); this provides a crucial control land-use configuration against which other future scenarios can be compared. The scenarios had the following key characteristics.

Scenario 0: Business as Usual

Under this scenario much abandoned land would be transformed, mainly into urban use. Some patches of new urban use are also created from pastures and shrubland, but in a smaller proportion. This urban expansion is mainly predicted for the mountainous area in the northern region, where housing demand is currently increasing. Areas under crops that are surrounded by urban areas, or forest and shrubland borders, are often converted to urban use. Cultivation of irrigated crops tends to move into uses such as forest, shrubland, and crops. Most of the crop mosaic patches in the southwest of the region are destroyed as urban use, pasture, and irrigated crops take over.

Scenario 1: Accelerated Urban Development

Scenario 1 (Fig. 3.52) simulated a perceived increase in demand for urban and residential development based on a forecast of continued population growth in the region in the immediate future, which suggested increased demand for housing. According to the regional government, an amendment of Land Law 9/2001 could lead to a nearly 18% increase in urban land cover over the years after 2006. Under this scenario the development of reserves and protected land in the region and strategic land-use management were also blocked. In the event this worrying scenario did not come to pass, because less than two years after this work was carried out the Spanish housing bubble burst following the 2007 economic crisis, and urban development has remained very limited ever since. This underlines the importance of scenario approaches to mitigate this kind of uncertainty in model projections to future dates.

Scenario 2: Protecting Key Agroecosystems

Scenario 2 (Fig. 3.52) was created in response to consultation with a group of stakeholders (including farmers, researchers, and planners in the regional government agricultural and rural development departments) who identified traditional

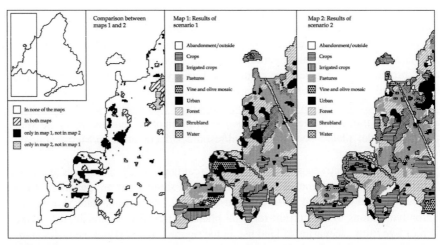

FIGURE 3.52

Outputs from the Madrid model redrawn in black and white from the original model output. Right and center: graphic representations of possible future land-use configurations for Scenario 1 (accelerated urban development) and Scenario 2 (protecting key agroecosystems). Left: map comparing the two scenarios.

farmlands and the strategic farming ecosystems as important sites for protection. The combination of olive trees, cereal crops, and vineyards, known as crop mosaics, is found in the southern part of the region and is an emblematic part of the landscape of this part of Madrid. The maintenance of these agroecosystems would sustain traditional landscapes and a characteristic structure of land ownership, where the average plot is 0.5 ha (INE, 2003) and land consolidation programmes have not been established. The management of these protected spaces would create belts of sustainable agriculture which would also encourage sustainable practices, improve educational and tourist activities, and act as limits to urban expansion. This is not as idealistic as it sounds, since multifunctional landscapes of this type were prioritized in subsequent rural develop programs (although the onset of the economic crisis in Spain shortly after this work was completed meant that funding did not materialize). Under this scenario, areas were created where intensive farming and urban development would be discouraged. Scenario 2 is thus similar to the baseline scenario in every respect except that a macro-scale constraint is imposed which encourages traditional agriculture and conserves these sensitive areas.

Scenario 3: Creating Buffer Zones and Belts Around Protected Areas

Scenario 3 simulates a proposal made by environmentalists to establish "ecological zones and zones of special use" to surround the natural protected areas. These areas would include sites protected by European regulations such as the EU Habitat Directive (92/43/EC) and EU Birds Directive (79/409/EC), and sites protected by national and regional legislation. A restraint on specific land uses covering a distance of 2 km will be established to prevent the influence of external pressures. Belts of sustainable agriculture would be created to mitigate the impact of intensive land use on the most fragile territory. In addition, this measure would favor the creation of ecological corridors between the protected areas.

Scenario 4: Riverbanks and Wildlife Corridors

This scenario responds to proposals to protect riverbank areas. Riversides and riverbanks have experienced the most drastic landscape changes in the whole of the region. The national Water Law (29/1985) establishes two restricted sections on each side of the main rivers: the first, "track land," is 5 m in width, and the second "vigilance area" is 100 m in width, where uses and activities are controlled. Nevertheless, the degradation of these areas suggests that a higher level of restriction may be necessary. Under this scenario, land uses were restricted inside two riverside buffer strips, one extending 200 m out from the river and a second strip extending from 200 to 400 m. All types of land use were constrained in the first 200 m buffer zone except forest. In the second 200 m buffer zone irrigated crops, abandonment, and urban categories are restricted. This two-tier buffer zone was established in parallel strips along the length of rivers and streams to represent the effects of targeted recuperation work.

Conclusions and Lessons Learnt

The five scenarios created with the Madrid policy simulator suggested a range of very different possible land-use futures, dependent to a large extent on the willingness of local planners to impose development restrictions and promote a multifunctional approach to land use in the region. Reviewing this work more than 10 years later, we can see that this did not happen largely because the economic crisis that plunged most of Europe into deep recession cut short all but the most urgent public initiatives. However, the utility of the scenario-based approach as an option generator for local communities and policy-makers can be clearly seen. With hindsight, we can see that it would have been useful to include an economic crisis scenario, where neither accelerated urbanization nor strategic planning is implemented, since this is what actually happened.

Under TiGrESS the dynamic geographical perspective of the research and the results of the conceptual model of land planning in Madrid that emerged as a result of a participatory process converged in the development of a land-use model. This contrasts with earlier developments using the same modeling framework, for example the SimLucia (White & Engelen, 1997) and The Monitoring Land Use/Cover Dynamics (MOLAND) (Engelen, Lavalle, Barredo, Meulen, & White, 2007) models, where the model was commissioned directly by policy-makers from the top down. However, a weakness of this project's approach is that the TiGrESS stakeholders' group, whose contributions helped to foster the model development, were not consulted about the results nor to what degree their expectations had been met. Also, while the stakeholders were involved in the framing of the problem that led to the model, they were not involved in the modeling process and only indirectly (through information given in interview sessions) in the scenario building. This was largely due to time constraints—the development of a simulation model

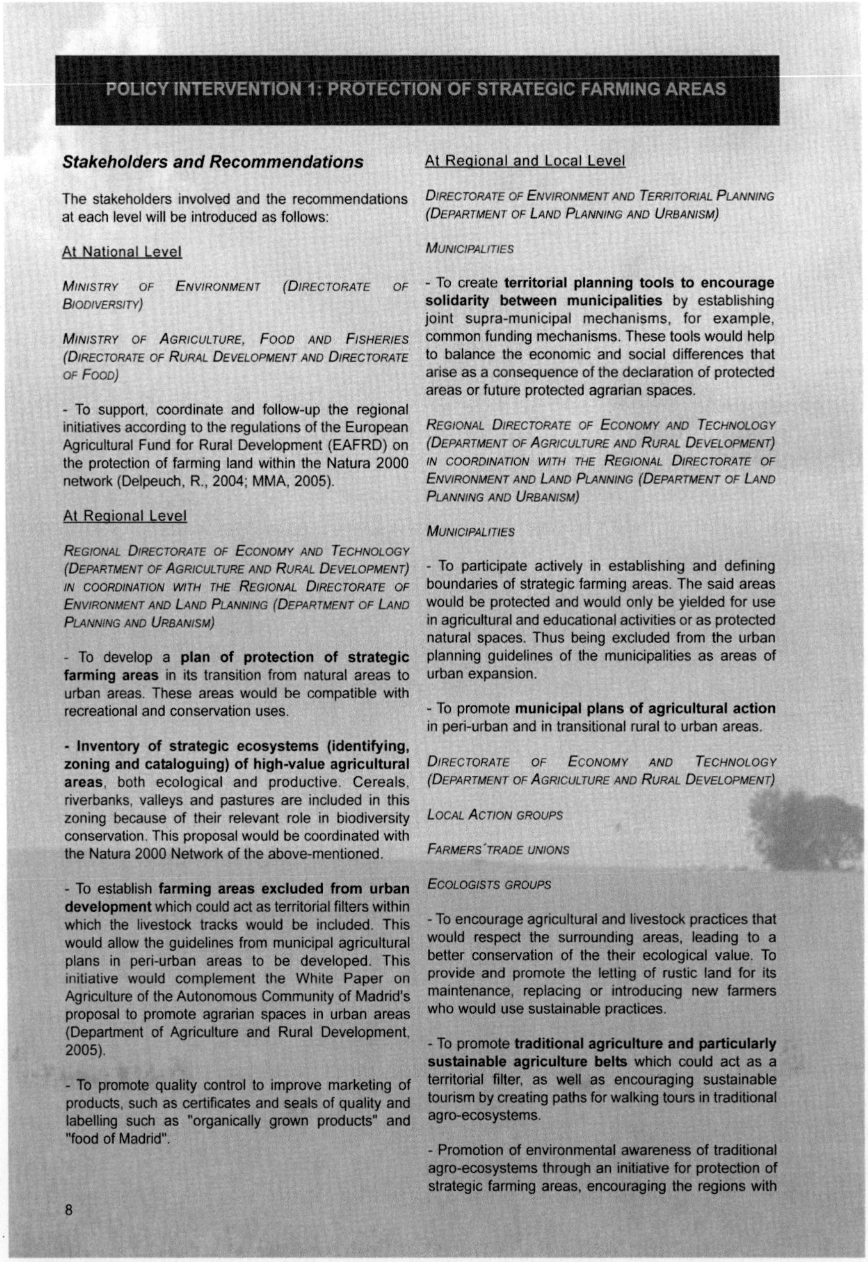

POLICY INTERVENTION 1: PROTECTION OF STRATEGIC FARMING AREAS

Stakeholders and Recommendations

The stakeholders involved and the recommendations at each level will be introduced as follows:

At National Level

MINISTRY OF ENVIRONMENT (DIRECTORATE OF BIODIVERSITY)

MINISTRY OF AGRICULTURE, FOOD AND FISHERIES (DIRECTORATE OF RURAL DEVELOPMENT AND DIRECTORATE OF FOOD)

- To support, coordinate and follow-up the regional initiatives according to the regulations of the European Agricultural Fund for Rural Development (EAFRD) on the protection of farming land within the Natura 2000 network (Delpeuch, R., 2004; MMA, 2005).

At Regional Level

REGIONAL DIRECTORATE OF ECONOMY AND TECHNOLOGY (DEPARTMENT OF AGRICULTURE AND RURAL DEVELOPMENT) IN COORDINATION WITH THE REGIONAL DIRECTORATE OF ENVIRONMENT AND LAND PLANNING (DEPARTMENT OF LAND PLANNING AND URBANISM)

- To develop a **plan of protection of strategic farming areas** in its transition from natural areas to urban areas. These areas would be compatible with recreational and conservation uses.

- **Inventory of strategic ecosystems (identifying, zoning and cataloguing) of high-value agricultural areas**, both ecological and productive. Cereals, riverbanks, valleys and pastures are included in this zoning because of their relevant role in biodiversity conservation. This proposal would be coordinated with the Natura 2000 Network of the above-mentioned.

- To establish **farming areas excluded from urban development** which could act as territorial filters within which the livestock tracks would be included. This would allow the guidelines from municipal agricultural plans in peri-urban areas to be developed. This initiative would complement the White Paper on Agriculture of the Autonomous Community of Madrid's proposal to promote agrarian spaces in urban areas (Department of Agriculture and Rural Development, 2005).

- To promote quality control to improve marketing of products, such as certificates and seals of quality and labelling such as "organically grown products" and "food of Madrid".

At Regional and Local Level

DIRECTORATE OF ENVIRONMENT AND TERRITORIAL PLANNING (DEPARTMENT OF LAND PLANNING AND URBANISM)

MUNICIPALITIES

- To create **territorial planning tools to encourage solidarity between municipalities** by establishing joint supra-municipal mechanisms, for example, common funding mechanisms. These tools would help to balance the economic and social differences that arise as a consequence of the declaration of protected areas or future protected agrarian spaces.

REGIONAL DIRECTORATE OF ECONOMY AND TECHNOLOGY (DEPARTMENT OF AGRICULTURE AND RURAL DEVELOPMENT) IN COORDINATION WITH THE REGIONAL DIRECTORATE OF ENVIRONMENT AND LAND PLANNING (DEPARTMENT OF LAND PLANNING AND URBANISM)

MUNICIPALITIES

- To participate actively in establishing and defining boundaries of strategic farming areas. The said areas would be protected and would only be yielded for use in agricultural and educational activities or as protected natural spaces. Thus being excluded from the urban planning guidelines of the municipalities as areas of urban expansion.

- To promote **municipal plans of agricultural action** in peri-urban and in transitional rural to urban areas.

DIRECTORATE OF ECONOMY AND TECHNOLOGY (DEPARTMENT OF AGRICULTURE AND RURAL DEVELOPMENT)

LOCAL ACTION GROUPS

FARMERS'TRADE UNIONS

ECOLOGISTS GROUPS

- To encourage agricultural and livestock practices that would respect the surrounding areas, leading to a better conservation of the their ecological value. To provide and promote the letting of rustic land for its maintenance, replacing or introducing new farmers who would use sustainable practices.

- To promote **traditional agriculture and particularly sustainable agriculture belts** which could act as a territorial filter, as well as encouraging sustainable tourism by creating paths for walking tours in traditional agro-ecosystems.

- Promotion of environmental awareness of traditional agro-ecosystems through an initiative for protection of strategic farming areas, encouraging the regions with

8

FIGURE 3.53

The guide, entitled "Toward Sustainable Land Planning in Madrid Region: Guidelines and Recommendations."

had not been envisaged in the project's original specification, and arose in response to real stakeholder needs at the end of the project.

The Madrid policy simulator developed under TiGrESS was intended to respond to bottom-up needs. Nevertheless, despite the strong participatory basis, the model was developed and calibrated by external experts. It remained, from the point of view of most stakeholders, a "black box." The Madrid policy simulator can be considered halfway between top-down policy-relevant models (like SimLucia or MOLAND) and completely participatory models (e.g., companion modeling approaches: Barreteau et al., 2003) in which virtually all modeling actions are carried out with stakeholders.

Thus when the opportunity arose to apply a similar methodology in the context of the DUSPANAC project (next example), the team included new participatory strategies to integrate this essential dimension.

3.3.1.2.4 Action Plans (Chapter 2, Section 2.1.2.3.6)

The extensive stakeholder engagement process culminated in the production of a concrete set of proposals in the form of a guide (Fig. 3.53, Table 3.8) that listed the policy options already explored (revised in the light of feedback from stakeholders). The aim of this guide was to begin the process of building a stakeholder network for participatory planning, linking a wide range of stakeholders in the land planning process across local and regional levels. Following a brief introduction to the current situation, the proposals and the expected responsibilities of the stakeholders involved were outlined. Seventy-five copies of this booklet were published in autumn 2005 (Encinas and Winder, 2005) and distributed to the key stakeholders (planners, farmers, and politicians) in the land planning process. The publication of this policy guide was timely, since the regional legislation (Land Law 9/2001) in force at that time was due to be replaced. Consequently, the extensive stakeholder engagement work carried out offered a unique chance to influence regional planning

Table 3.8 Summary of the Proposal of Actions for Each Action Group Involved, and for the Problems Identified

Action Number	Action Proposed
1	Protection of traditional farmland
2	Support for the development of agroecology
3	Promotion of local markets
4	Recuperation of riverbanks
5	Improvement of access to and availability and quality of environmental information
6	Coordination between environment, agriculture, and planning domains
7	Promotion of participatory processes

processes over future land-use schemes and guidelines to apply in the region of Madrid to tackle the deficiencies in participatory processes.

3.3.2 CASE 3.3.2: DUSPANAC: MODELING LAND-USE DYNAMICS IN THE SPANISH NETWORK OF NATIONAL PARKS AND THEIR HINTERLANDS[33] (2011−2013)

3.3.2.1 Project Synthesis

Growing human pressure on the biosphere has driven scientific activity in many different disciplines over the last three decades. Changes in both the physical characteristics of the Earth's surface (land cover) and its exploitation by humans (land use) are known to affect a wide range of Earth systems, like soil, water, climate, and ecosystems. Analysis and monitoring of land use and land cover change (LUCC) is therefore essential to understanding and managing global change.

LUCC is naturally at its most intense where human activity is most densely concentrated, yet even protected areas remote from permanent human occupation, such as national parks, may experience serious land-use changes, although to a lesser degree than other areas without such protection. Human activity, climate change, and the dynamics of the ecosystems present inside the parks provoke changes in landscape morphology and the mosaic of land-use categories, compromising the system resilience of the parks and their surroundings. Accurate appraisal of the present situation and future prospects in these important natural areas, coupled with a process of engagement with relevant stakeholders, is indispensable for their good management and the adequate conservation of the values that make them worthy of the protection they enjoy. The DUSPANAC project (http://www.geogra.uah.es/duspanac/; Escobar, Hewitt, & Hernández Jiménez, 2016), which began in 2011, looked to address this challenge for Spain's 14 national parks.[34] The project had the following aims.

- To characterize, quantify and understand land-use changes in the Spanish network of national parks and their hinterlands since 1990.
- To carry out modeling and mapping of land-use dynamics and project these dynamics to a 25−30-year horizon through different scenarios based on environmental conditions (climate change among others) and sociopolitical restrictions relating to the management and use of the parks.
- To establish participatory processes with the key national parks network stakeholders and target population to permit the appropriate development and

---

[33]Funded by the OAPN, a division of the Spanish Environment Ministry, under the 2010 Programme for Scientific Research Projects in the Network of National Parks. Project directed by the University of Alcalá, Madrid, Spain (project coordinator Francisco Escobar).

[34]In the final stages of this project a new national park was declared in the Sierra de Guadarrama, Madrid region.

implementation of decision-support tools in the management and conservation of the parks and their hinterlands.

The project was developed in three main phases.

1. Geographical analysis in Geographical Information Systems (GIS) of LUCC since 1990 in all 14 national parks to quantify LUCC, understand LUCC trajectories, and identify the parks at greater risk of ecological degradation on the basis of LUCC speed, amount, and significance (Hewitt, Pera, & Escobar, 2016).
2. Select the most threatened national park on the basis of this analysis and develop a land change modeling exercise in that location with key stakeholders. The intention was to integrate the stakeholder community as far as possible in the modeling process, building on experience from earlier projects such as Tigress, to build a fully participatory (co-developed) land-use model (Hewitt, Van Delden, & Escobar, 2014).
3. Use the model to develop a range of future scenarios, evaluate the model performance, and develop a battery of indicators for future management of the park—all of these activities to be undertaken by the stakeholder group (Hewitt, Jiménez, Navarro, de la Cruz Lecanda, & Escobar, 2016).

The results of the first phase indicated that park that had seen the most extensive, fastest, and most diverse LUCC by far was Doñana, an internationally recognized area of importance for biodiversity located in the Guadalquivir River estuary in southwestern Spain. This area was selected for in-depth research in phases 2 and 3, arising from the question: "Why, given the recognized importance of the Doñana natural area and the many national and international protection designations it has enjoyed for many years (Biosphere Reserve, UNESCO World Heritage property, national park, natural park, Ramsar site, etc.), is it still at risk of degradation inferred by the observed LUCC?"

This question was the starting point for an in-depth process of participatory research and modeling, which is described below.

3.3.2.2 Description of Techniques
3.3.2.2.1 Geographical Analysis of the Territory (Chapter 2, Section 2.1.2.2.7)

Prior to beginning the stakeholder engagement process, a detailed analysis of LUCC in the network of national parks was undertaken (Fig. 3.54). The analysis was carried out with the free open-source GIS software ILWIS (http://52north.org/communities/ilwis/ilwis-open/download) for two periods, 1990−2000 and 2000−2006 using the freely available Corine Land Cover (CLC) database (http://land.copernicus.eu/pan-european/corine-land-cover). Significant LUCC was observed for Cabañeros, Doñana, Garajonay, Monfragüe, Picos de Europa, Sierra Nevada, and Tablas de Daimiel national parks. In all cases transitions were observed across CLC level 1 categories, artificial (level 1, group 1), agricultural (level 1, group 2), natural (level 1, group 3), and

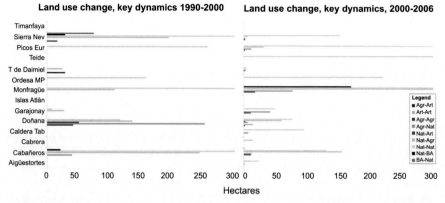

FIGURE 3.54

(A) (left) Key change dynamics in Spanish national parks, 1990–2000. (B) (right) Key change dynamics in Spanish national parks, 2000–2006. Legend: Art—artificial surfaces (e.g., urban fabric, leisure areas); Agric—agricultural areas (e.g., non-irrigated or irrigated); Nat—natural areas (e.g., shrubs, forests); ZQ—burnt areas (e.g., as a consequence of forest fire).

wetlands (level 1, group 4) as well as within these categories.[35] Of all of the 14 parks considered by the analysis, Doñana was outstanding in terms of LUCC, combining a high rate of change, change in both the time periods analyzed, and high variability of change between different groups of land uses (artificial, agricultural, natural, wetlands). The combination of urban development and expansion of cropland under irrigation in period 1 (1990–2000) around the fringes of the protected area, and the very diverse combination of different human activity systems in close proximity to the most important conservation zones led us to identify Doñana as the most threatened of the 14 national parks, at least by this measure (Hewitt, Pera et al., 2016).

3.3.2.2.2 Stakeholder Mapping and Identification of Key Informants (Chapter 2, Section 2.1.2.1.1)

On the basis of the results of the LUCC analysis discussed above it was decided to progress to phase 2, the development of a land-use model to simulate possible future land-use change for the study area of Doñana. The DUSPANAC researchers began to review recent and ongoing environmental research in the area to gain a better understanding of the task that faced them. This process led them rapidly to another group of Madrid-based researchers who were already working in Doñana. This team[36]

---

[35]For example, transition from natural to agriculture (across CLC level 1 categories) and from scrubland to woodland (within CLC level 1 category 3, "natural").

[36]We gratefully acknowledge the assistance provided at all stages of the project by this research team from the social ecological systems laboratory of Madrid Autonomous University, without whom our work would have been considerably less successful.

kindly invited us to participate in one of their stakeholder workshops on mapping of ecosystem services. In this way we were able to learn directly from these researchers about Doñana and its particular set of problems, and to present our proposed work to the existing stakeholder group; this certainly enhanced our credibility and facilitated our subsequent interactions with the stakeholders. The team also generously shared with us their extensive list of contacts. Though there is no doubt that this was a piece of luck which we could not have prepared for, it does serve to illustrate the great importance of finding out about other developments or initiatives that may be related to the proposed project, as well as the importance of key informants who can sometimes (as in this case) play a transformative role.

Stakeholders who participated in the workshops are listed in Table 3.9.

3.3.2.2.3 Participatory Modeling (Chapter 2, Section 2.1.2.3.1)

To build the land-use model we used an updated version of the Metronamica software, described earlier for the Tigress project, to simulate growth, pressure, and competition between different land uses under a range of hypothetical future scenarios. However, as a result of our learning experiences in TiGrESS we were able to make significant advances in the participatory modeling method we used.[37] Under Tigress, the land-use model was developed at the end of the process, in response to the need for specific information identified by the participant group, but the stakeholders were not involved in the modeling process itself and there was no opportunity to share the finished model with them or collect their feedback. Under Duspanac the modeling objective was a key part of the project proposal, so its development could be initiated early on. This provided an opportunity to co-develop the model with the stakeholder group, involving them at all stages of the project. The group worked on a wide range of activities carried out over three participatory workshops (WS), from the identification of land-use categories and land change trends (WS1), through model calibration and scenario development (WS2), to participatory evaluation of the model and the modeling process itself (WS3) (Hewitt, Hernández Jiménez, Román Bermejo, & Escobar, 2017). The time between workshops was used by the research team to work on non-participatory aspects of the model, such as the statistical goodness-of-fit testing of calibrations, which were communicated to stakeholders at each workshop. We thus conceived of the model as a sequence of analytical and discursive activities by which the knowledge of both the research team and the participant group was gradually increased (Fig. 3.55). We note that although this figure presents the model as a sequence of activities, with clear start and end points, the process can also be seen as a *knowledge cycle*: although the knowledge state of the model system is not the same at the end as at the start of the process (new knowledge has undoubtedly been created), the new state gives rise to new questions, which may lead to further iterations of the process.

---

[37]This is reported in Hewitt et al. (2014) and Hewitt et al. (2017).

Table 3.9 Workshop Participants by Sector. In Addition, Three or Four Researchers Attended Each Workshop as Facilitators

Stakeholder, by Sector	WS1	WS2	WS3
Science			
Researcher, Autonomous University of Madrid	yes	yes	yes
Researcher and university lecturer, University of Seville	yes	yes	yes
Researcher, Doñana Biological Station (national scientific institute)	yes	yes	yes
Agriculture			
Director, Federation of Rice Farmers, Seville	yes	yes	yes
Representative, Young Farmers' Agricultural Association (ASAJA)	yes	yes	yes
Representative, Andalusian Farmers and Livestock Keepers Union, Huelva division	no	yes	yes
Tourism			
Tourism representative, Doñana natural area	no	yes	yes
Local policy-makers			
Moguer Municipal Council, environment technician	yes	no	yes
Representative, Doñana 21 Foundation	yes	yes	yes
Almonte Municipal Council, environment technician	no	no	yes
Regional policy-makers			
Regional administration, environmental research division	no	yes	yes
Regional administration, environmental research division	no	yes	yes
Natural area managers			
OAPN, head of project monitoring	yes	yes	no
Director, Doñana Natural Area	yes	yes	yes
Subdirector, Doñana Natural Area	yes	no	no
Director of conservation, Doñana Natural Area	yes	no	no
Director of public use, Doñana Natural Area	yes	yes	yes
Guide, Doñana Natural Area	no	no	yes
Monitoring division, Doñana Natural Area	yes	yes	yes
Environmentalists			
Ex-Ecologistas en acción (environmental group)	yes	no	yes
World Wildlife Fund	no	no	yes

Procedure for development of an integrated particpatory/analytical land use model

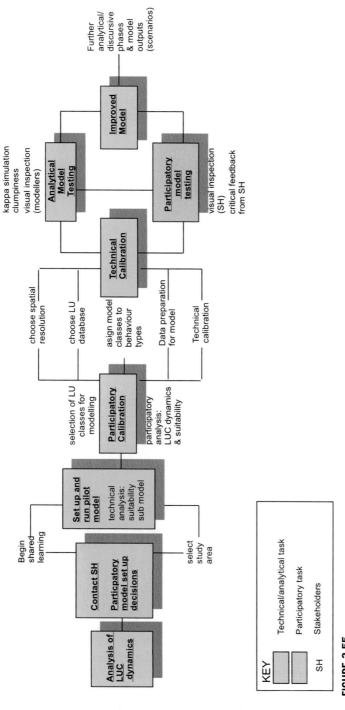

FIGURE 3.55

The Doñana land-use model as a sequence of alternating participatory and non-participatory activities.

This integration of these participatory aspects into the land-use modeling had two main advantages. On the one hand, the model that was developed is demonstrably "a better model" than it would otherwise have been, in the sense that it responds to the real concerns of the stakeholders about the study area and is not just the researchers' impression of the problem. On the other hand, we addressed the thorny issue of usability (most so-called decision-support or policy-support tools are never actually used by anyone other than the developer), since the participants not only helped to build the model by supplying data and criticizing its performance, but actually sat in front of the computer and used it to run simulations. Thus although the model was not used afterward in the day-to-day work of the stakeholders (e.g., for local planning or conservation area management), the stakeholder group did learn new ways of looking at the problem which, very likely, have since become embedded in their approach. That the workshop participants did in fact gain useful new knowledge from their participation in the modeling process is shown by their responses to the evaluation. And of course, we, the researchers, learnt a great deal about the study area itself, about group dynamics, about the roles of the different sectors and actor groups in the natural area, and about our own model as we struggled to adapt it to meet the concerns and aspirations of the participants. The participatory modeling approach employed led, we believe, to what Reed et al. (2010) refer to as "a change in understanding that goes beyond the individual to become situated within wider social units or communities of practice through social interactions between actors within social networks." This change is known as social learning.

3.3.2.2.4 Discussion Groups (Chapter 2, Section 2.1.2.2.3)

To elicit the information required for the model we carried out a range of activities in group discussions with the workshop participants. For most of the discussion group work, participants were divided into three groups of 4 or 5 (each workshop had 12—14 participants) After working together for an initial period, each group presented its conclusions in a plenary session. In many cases, groups were also asked to complete activity sheets.

Because of the close focus on modeling patterns of land-use change in the area, the activities we carried out were quite specific. In the following paragraphs we describe a few of the most important of these.

In the first workshop, held on February 22, 2012, participants undertook three specific activities essential to set the basic parameters of the model: classification of land-use categories for use in the model; analysis of land-use dynamics and drivers of change in Doñana; and classification of suitability of the study area to be occupied by different land-use categories.

3.3.2.2.4.1 Classification of Land-Use Categories for Use in the Model

In the first activity participants were tasked with defining the most relevant land-use categories to explain the dynamics of change in Doñana from their own knowledge and experience out of a much larger set of categories. Prior to the workshop the research team had undertaken a preliminary reclassification of the Andalusian

FIGURE 3.56

Participants working together to agree on land-use categories for the spatial model.

regional government land-use and vegetation cover database,[38] reducing 107 land-use categories to 48. This was still too many for a workable land-use model, but further aggregation clearly required local knowledge. Participants were tasked with building a new, simpler land-use scheme to represent synthetically and as realistically as possible key land-use dynamics from the 48 starting categories. A set of cards illustrating and describing the 48 categories was handed out to each group to support their discussion. Each group chose a spokesperson to present their categorization to all participants. Finally, a common and representative "consensus" classification was obtained through general discussion with all participants (Fig. 3.56). The tangible result of this first activity was a definition of the final 23 categories of land uses that would be introduced in the model.

3.3.2.2.4.2 Analysis of Land Use Dynamics and Drivers of Change in Doñana

In the second activity stakeholders were tasked with conducting a critical analysis of the land-use changes detected in Doñana through the LUCC analysis undertaken at the beginning of the project. Researchers circulated nine maps showing different types of land-use change from the CLC database for two periods, 1990−2000 and 2000−2007. The nine land-use dynamics analyzed were as follows.

---

[38]REDIAM: Mapa de usos y coberturas vegetales del suelo de Andalucía [1:25,000 scale map of land use and vegetation cover for Andalucia] (MUCVA) 1956−1977−1984−1999−2003−2007, escala 1: 25.000. Consejería de Medio Ambiente y Ordenación del Territorio. Junta de Andalucía. Available at http://www.juntadeandalucia.es/medioambiente/site/rediam/menuitem.04dc44281e5d53cf8ca 78ca731525ea0/?vgnextoid=ca74d2aa40504210VgnVCM1000001325 e50aRCRD&vgnextchannel=7b3ba7215670f210VgnVCM1000001325e50aRCRD& vgnextfmt=rediam.

1. Loss of natural areas.
2. Increase of natural areas.
3. Increase in artificial surfaces.
4. Increase in irrigated crops.
5. Increase in pasture and dryland crops.
6. Change from shrubland into woodland (all types).
7. Change from woodland to shrubland (all types).
8. Changes (losses and gains) to wetlands and marshlands.
9. Burned areas.

Each group responded to a series of questions about these land-use dynamics contained in a pro-forma worksheet (Fig. 3.57).

As in the previous activity, participants worked in groups and later shared their conclusions with the whole company. This activity gave the research team a much better understanding of the land-use dynamics of the study area and enabled the model transition rules (neighborhood, zoning, and suitability) to be defined, allowing the model calibration to begin.

3.3.2.2.4.3 Classification of Landscape Suitability

In a third activity, participants evaluated the influence of particular suitability factors (rainfall, slope, temperature, etc.) on each land-use class in the study area through a simple qualitative scoring system: strong (mucho), weak (poco), or no influence at all (nada). On the basis of this information, an agreement or confidence index

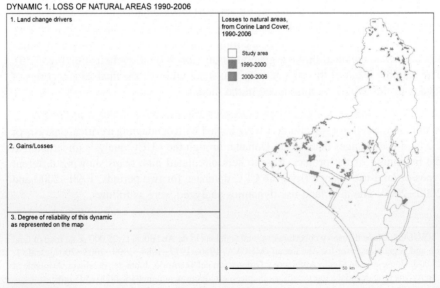

FIGURE 3.57

Pro-forma worksheet for analyzing land-use dynamics.

(C) was calculated by allocating a value of 0 where all three groups disagreed, a value of 1 where two groups disagreed with the third group, and 2 where all groups agreed. These values were then summed to give a total agreement index for each suitability factor. The categorical responses (strong, weak, and no influence) given by the stakeholders for each land use against a given factor were translated into a simple scoring system, referred to here as the influence index (I) of 2 (strong), 1 (little), and 0 (no influence). Finally, the confidence index (C) was multiplied by the influence index (I) to give a total overall score by land use for each suitability factor. Thus, for example, in assessing the PLASTIC (intensive crops under plastic) land use, all three groups felt slope to be an important influence and noted a "strong" score of 2, giving $(2 + 2 + 2) = 6$. Since all groups were agreed about the importance of slope for this land use, the highest confidence score (2) was allocated. Thus the total score for the slope factor for the PLASTIC land use was 12 (6×2), indicating that the stakeholders felt with a high degree of confidence that slope was influential in determining the location of crops under plastic, lesser slopes being preferred. Suitability parameter settings inside the model were estimated on the basis of this information. For example, in the case of the classes of industrial (IND), rice (RICE), intensive crops under plastic (PLASTIC), and intensive woody crops (INTWOOD), high suitability parameter values were given to areas with slopes of less than 5%.

3.3.2.2.5 Evaluation of Model Outputs (Chapter 2, Section 2.1.2.4.1)
3.3.2.2.5.1 Participatory Assessment of Model Goodness-of-Fit
In the second workshop, held on December 11, 2012, participants worked with the computer simulation model directly. After a brief explanation and introduction, participants were tasked with evaluating model goodness of fit (Fig. 3.58): they were divided into three groups and given four simulations representing different stages

FIGURE 3.58

Participants analyzing the goodness-of-fit of various model calibration runs.

Table 3.10 Calibration Milestones and Their Relationship to Individual Simulations Evaluated by Stakeholders in Bold

Milestone Number	Simulation Run	Calibration Substep
1	1	Simple neighborhood rules only (benchmark model)
2	5	Calibrate neighborhood rules
3	**11**	**Add accessibility**
4	21	Add suitability
5	**23**	**Add zoning**
6	**34**	**Adjust suitability parameters**
7	**35**	**Adjust neighborhood parameters**

of the calibration process (Table 3.10), though only two simulations (11 and 35) were successfully evaluated by all participants in the time available. Simulations were provided both on-screen and as paper printouts. Participants explored the simulations and debated the merits of each in groups. Subsequently they rated the similarity of location and degree of clustering of simulated land-use categories compared to real land-use categories on a pro-forma worksheet. This activity aimed, first of all, to acquaint the stakeholders better with the detailed process of creating land-use simulations and remove a little of the mystery surrounding the operation of the model. But the activity also aimed to enhance the validity of the process of visual inspection of calibrations, something researchers normally undertake themselves. Clearly, while evaluation of goodness of fit with the human eye is highly subjective, subjectivity decreases when simulations are evaluated by many people.

3.3.2.2.6 Scenario Development (Chapter 2, Section 2.1.2.3.3) and Participatory Mapping (Chapter 2, Section 2.1.2.3.2.2)

Following the participatory assessment of model goodness-of-fit, participants worked directly with the model on two key tasks: estimation of land-use demand for four model scenarios adapted from the Doñana eco-futures (see Palomo, 2012; Palomo, Martín-López, López-Santiago, & Montes, 2011 for a full description of the eco-future scenario development process), and estimation of new land-use location through a participatory mapping exercise.

3.3.2.2.6.1 Estimation of Land-Use Demand for the Eco-Future Scenarios

The next stage of the process was to use the spatial model to develop land-use projections for future dates, to represent graphically some possible ways in which the protected area might evolve in response to particular conditions (scenarios). We chose to use a set of pre-existing scenarios which local stakeholders, including many of the participants present in our workshops, had recently developed with the previously mentioned team of social ecosystems researchers from Madrid Autonomous University. There were four eco-future scenarios; *Doñana globalized* knowledge, *Trademark Doñana*, *Arid Doñana*, and *Adaptive Doñana: wet and wild*, representing a broad range of ways in which stakeholders imagined that the territory might evolve.

However, these scenarios were not suitable in their existing form for direct inclusion in the model, since the land-use categories selected by stakeholders in WS1 and later used to build the model were not explicitly present in the scenario narratives. For new land to be allocated in the model, it is necessary to define the amount of new land to be allocated, known as the *demand*, for each land use. In urban modeling applications it is common to use projected population growth to estimate the amount of new land required at each time step (see e.g., Hewitt & Díaz Pacheco, 2017). In this case we wanted to link the patterns of change expressed in the scenario narratives with a quantifiable demand for different types of land. For example, in a climate change scenario in which water becomes scarce (e.g., Scenario 3: Arid Doñana), it might be expected that demand for water-intensive crops would be reduced.

Thus the participants' first task was to analyze the scenario narratives looking for references to land-use change, which they then related to a land-use category. On this basis stakeholders estimated land-use demand for the model for each category under each scenario, and drew a tendency line showing how the expected land use would develop to the scenario end date, defined as 2035. The participants gave detailed explanations for their decisions in each case, shown in Table 3.11. To the four eco-future scenarios analyzed by the participants, we added a fifth scenario, Scenario 0, business as usual, in which demand for each land-use category was obtained simply by continuing the existing trends identified from the historic patterns of change identified in the territory.

With the demand estimated for each of these scenarios, it was now possible to run the model to 2035, allowing it to allocate the number of cells represented by the demand for each land use under each scenario in the highest potential areas (Fig. 3.59).

Finally, although the model was now ready to be used to generate land-use maps for each scenario in 2035, there was one last important task for the participants. We wanted to see whether the allocation process in the spatial model, which responded to the transition rules developed during the calibration process, actually corresponded to those areas in which stakeholders expected change to happen. So in the final activity of the workshop, stakeholders located the land-use demand they had estimated in the previous activity using colored buttons representing different quantities of land (e.g., large button 500 ha, small button 100 ha) on an A0 paper plot of the 2007 land-use map (Fig. 3.60), including an explanatory legend where necessary. Results were recorded photographically. Coordinates were included on the map for subsequent georeferencing in GIS software.

---

[39]These uses were modeled as *passive*, as they grew or declined in response to the behavior of the *active* land uses, whose surface area is assigned in the model according to the hectare counts given in this table. For this reason, they do not appear in the table.

Table 3.11 Land-Use Demand Estimation and Participants' Explanations of Land Change Tendencies for Each Eco-Future Scenario up to 2035

| Active Land Use | Eco-Future Scenario | | | |
	Scenario 1: Doñana Globalized Knowledge (Group 1)	Scenario 2: Trademark Doñana (Group 2)	Scenario 3: Arid Doñana (Group 3)	Scenario 4: Adaptive Doñana: Wet and Wild (Group 4).
Urban	6200 ha Lower than the amounts defined under the linear trend scenario	12,500 ha Considerably higher than the linear trend scenario	Number of hectares similar to 2007 due to a tendency toward depopulation and migration to large cities	Considerable decline in urban land due to recovery of the coastal strip and demolition of all abandoned urban land
Industrial	1999 ha Similar to the linear trend scenario	2500 ha Higher than the linear trend scenario	Increase in industrial land due to expansion of sustainable energy	Considerable decline from the linear trend scenario down to 1999 levels
Rice	15,410 ha Lower than the linear trend scenario, similar to 2007 levels	15,400 ha Reduction in surface area down to 2007 levels.	Would disappear—very different to the linear trend scenario	Considerable decline, reaching 1965 levels—approximately 2000 ha
Crops under plastic	6000 ha Lower than the linear trend scenario, similar to 2003 levels	12,500 ha Considerably higher than the linear trend scenario	They remain the same and even begin to decrease due to increased production costs	Decline in land occupied by intensive crops in general, giving way to allotments and market gardens
Woody irrigated crops	8633 ha Lower than the linear trend scenario, similar to 2007 levels	4700 ha Sharp decline to 1999 levels	Sharp decline	Decline
Other irrigated crops	26,392 ha Lower than the linear trend scenario, similar to 2007 levels	7000 ha Sharp decline down to 1956 levels	Sharp decline	Decline

Non-irrigated crops	62,000 ha Higher than the linear trend scenario	70000 ha Higher than the linear trend scenario, back up to 1999 levels	Sharp increase, but with problems of salinization appearing by 2035	Increase to approximately 70,000 ha, close to 1999 levels
Vine and olive	33,422 ha Higher than the linear trend scenario, similar to 2007 levels	15,000 ha Sharp decline with respect to the linear trend scenario	Slow increase in surface area	Considerable increase to 45,000 ha
Eucalyptus	2000 ha Considerable decline	12,000 ha Considerable increase in surface area occupied by this category owing to its use as an energy crop	Considerable increase in surface area occupied by this category owing to its use as an energy crop	Decline, since this is an invasive species
Coniferous woodland	29,914 ha Lower than the linear trend scenario, similar to 2007 levels	15,000 ha Considerable decline in surface area	Surface area occupied by coniferous shrubs would increase	Would remain the same
Mixed and other woodland	7500 ha Considerable increase, taking over the land formerly occupied by Eucalypts	1000 ha Similar to decline observed under the linear trend scenario	Decline to an equilibrium state	Decline, reaching 1956 levels
Remarks	Some land uses are missing—marshland, dunes, etc.[39]	Missing natural and water areas	A prolonged drought would bring the Doñana natural area to a point of complete instability in a short space of time	Subsidiary scenario: missing uses, wetlands, river banks, Mediterranean wild land, grasslands, coastal, inland waters, market gardens[38]

Scenario 1 (2035): Doñana globalized knowledge

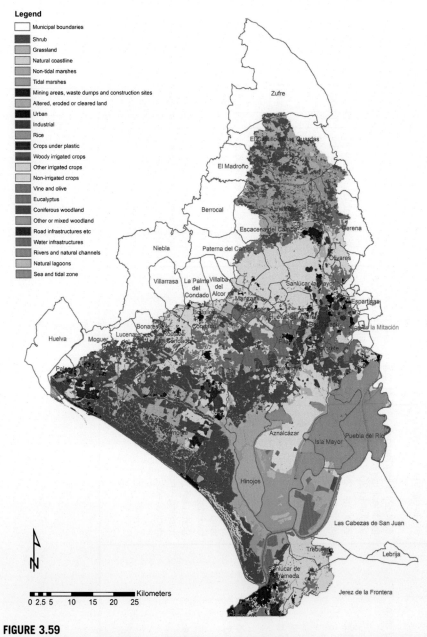

FIGURE 3.59

Scenario 1: Doñana globalized knowledge.

FIGURE 3.60

Demand estimation and land-use allocation for Scenario 4: Adaptive Doñana, wet and wild.

3.3.2.2.7 Participatory Process Evaluation (Chapter 2, Section 2.1.2.4), Dartboard Technique (Chapter 2, Section 2.1.2.4.3), and Participation Ladder (Chapter 2, Section 2.1.2.4.2)

The last project workshop, held on September 25, 2013, was mainly dedicated to evaluating the model and the participatory process. Before undertaking the evaluation exercise, participants explored the future land-use simulations that had been generated for each of the five scenarios (business as usual, a linear extrapolation of past land-use tendencies, plus the four eco-future scenarios discussed previously). They noted down their impressions and then discussed them in the plenary session.

The participatory process was evaluated through three linked activities.

1. Rapid-response questionnaire about the workshop process.
2. Dartboard technique to identify activities that participants found most successful for collective discussion and reflection, most difficult, and most relevant to the different actors involved in Doñana.
3. Locating the process on the participation ladder.

3.3.2.2.7.1 Rapid-Response Questionnaire

Stakeholders were invited to respond to two questions; has your knowledge about change in the Doñana natural area increased through participation in the workshops? Have you had any new reflections about the future of the Doñana natural area as a result of your participation in the workshops? In the case of affirmative responses, participants wrote their new knowledge/new reflections on post-it notes and added them to the wallchart (Table 3.12).

3.3.2.2.7.2 Dartboard Technique

In the second part of the evaluation activity, participants were asked to rate the individual activities of the participatory process and its applicability to the stakeholder community through the dartboard technique (Heras López, 2015; O'Brien & Moules, 2007; WAC, 2003) by responding to the following questions.

Table 3.12 Stakeholders' Own Evaluation (Verbatim Transcript in Translation) of New Knowledge and New Reflections Acquired as a Result of the Process

Utility of New Knowledge Acquired From the Process	New Reflections Arising as a Result of the Process
To apply the data to reports and studies which were not previously available	Land change drivers
Training/learning	Importance of prioritizing sustainable use of water in general
Awareness of the complexity of the factors that influence the predictability of the new scenarios	Coastal erosion will endanger urban developments
To improve my habitual work activities	Rice and intensive crops would be reduced under climate change
To prevent situations of environmental degradation	Consequences arising from soil sealing
	I feel that I participate in the future but I do not think that others feel that same way
	The trends in most cases indicate a difficult future for the natural area

- In your opinion, which activity has been most successful at facilitating collective discussion and reflection?
- Which activity did you find the most difficult?
- How applicable and relevant is the model for the different stakeholders?

First, a reference document describing the different activities carried out, including pictures, was handed out to the participants as an *aide memoire*. Three dartboards were drawn on the wall (Fig. 3.61). The first two dartboards were each divided into six wedges, with each wedge corresponding to an activity to be evaluated. The third dartboard was divided into ten sectors representing the stakeholder community (Fig. 3.62). The participants were invited to approach the dartboards and respond individually to the questions posed by drawing a dot in the dartboard, such that the closer to the center of the dartboard, the better the score. Once the scoring was complete, a common reading of the results was made, grouping together the positions, and the results were discussed and analyzed by all participants in a plenary session. The main findings from the discussion were written down on the wallchart.

The land-use classification exercise (WS1, Activity 1) and the participatory analysis of land-use dynamics (WS1, Activity 2) were highly rated by participants for group reflection and discussion. Stakeholders regarded the participatory evaluation of model goodness of fit (WS2, Activity 1) as the most difficult of the activities undertaken. For the third question, most participants rated the model's relevance and applicability highly for most sectors (Fig. 3.62), but the scatter of points on the outer circle for all sectors except regional land planning and farming shows that some participants remained skeptical about the model's usefulness in general.

FIGURE 3.61

Participants working on the dartboards.

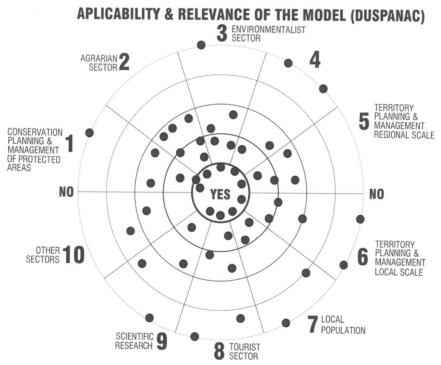

FIGURE 3.62

Graphic representation of results for question 3: how applicable and relevant is the model for the different stakeholders?

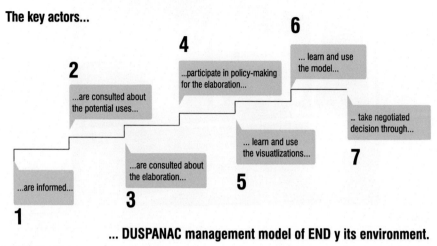

FIGURE 3.63

Participation ladder regarding key stakeholders' involvement during the modeling process in the DUSPANAC project.

3.3.2.2.7.3 Participation Ladder

One of the main aims of the Duspanac project was to involve key stakeholders in modeling land-use changes in the Doñana natural area and its hinterland to support territorial decision-making. To assess the degree of success achieved regarding this objective, a participation ladder adapted to the project was created (Fig. 3.63). We wanted to know to what extent stakeholders had felt involved in the process. Participants were divided in three groups, each of which, following group discussion, had to draw a cross indicating which step they felt had been reached in the participatory process. Groups 1 and 2 located the process on the third stair from the top, while group 3 located it a step below, since they felt that they had not had sufficient time to learn fully how to use the model outputs.

3.3.2.2.8 SWOT Matrix (Chapter 2, Section 2.1.2.1.7)

Analysis of SWOT is often carried out at the appraisal stage of a project to arrive at a rapid initial diagnosis of the issues that are going to be studied. In the DUSPANAC project SWOT analysis was used by the researchers at the evaluation stage, to understand the problems and potential of the model as a decision-support tool for use in planning of the natural area. The analysis was undertaken by five researchers shortly after completion of the last of the three workshops, in the light of the results of the various evaluation exercises carried out by the participant group. Table 3.13 shows the results of this exercise.

The SWOT analysis was useful for generating insights into the process, understanding which aspects had been most successful, and, most important of all, planning future actions to give continuity to the activities developed here. One of these, *O4. Translate model to a free software environment*, was a challenge we addressed under COMPLEX, the next project described.

Table 3.13 SWOT Matrix, DUSPANAC Project

Strengths	Weaknesses
S1. The model scale is appropriate for regional decision-making (water, protected areas)—W1	W1. The scale of the model is not appropriate for municipal decision-making (regional model)—S1, O8
S2. Training/social learning has taken place, both the research team and the stakeholders have gained new knowledge	W2. Not all the relevant stakeholders were made aware of the importance of their knowledge for the model development—T2
S3. New ideas for application of the model were generated	W3. Commercial software (expensive, limited access)—T1
S4. Model and software visualizations are very attractive—O4	W4. Interests of some stakeholders were not represented (e.g., crop prices and other economic factors)—T1, T2, T3
S5. Stakeholders' knowledge has been directly included in the model	W5. Water management not present
S6. Stakeholders see their contributions reflected in the model	W6. No user manual available for the visualization software (The Map Comparison Kit (MCK))
S7. Work has been closely linked to a previous research project—O2, O3	W7. Land use and ES were not linked—O3
S8. Model employs well-known commercial software (technical support, users community, well tested)—W3	W8. Excessive emphasis on environmental sector—O8
S9. Joint reflection spaces have been created	W9. Lack of interest and participation from planning decision-makers (local authorities)
S10. Complex revealed as simple	
Opportunities	**Threats**
O1. Stakeholders can use all the model results, available on the website	T1. May not be used (model, visualizations, results, etc.)
O2. Development of future projects/collaborations with stakeholders involved in the process	T2. May be seen as an academic tool
O3. Potential to link land use and ES	T3. If the usefulness of the model is not clear to people involved, they may be discouraged from participating in further processes
O4. Translate model to a free software environment	T4. Scenarios are seen as very unrealistic by some people
O5. Stakeholders could model their own scenarios and indicators, supported by the research team	
O6. Development of new ideas that have emerged (see strengths)	
O7. Workshops or training courses based on the process, results, and tools	
O8. Keep working on identified weaknesses (local scale, non-environmental sector)—W1, W8	

3.3.3 CASE 3.3.3: COMPLEX: KNOWLEDGE-BASED CLIMATE MITIGATION SYSTEMS FOR A LOW-CARBON ECONOMY[40] (2012–2016)

3.3.3.1 Project Synthesis

Climate change is one of the greatest threats that human society is currently facing, with catastrophic degradation of Earth systems leading to potentially disastrous consequences for billions of people. In Europe, alongside the Paris climate agreement, ratified by the EU in 2016, hopes have been placed on a policy framework known as the Low-Carbon Roadmap, (LCR), in force since 2009. The LCR sets clear targets for reduced greenhouse gas emissions for 2020, 2030, and 2050 (see http://ec.europa.eu/clima/policies/strategies/2050/index_es.htm), and these will have to be met or surpassed to stand a chance of bringing global temperatures down in the long term. Unfortunately many member states, for example the Netherlands, the United Kingdom, and Spain (the latter being the subject of our case study), are struggling to meet these targets. One of the key goals of the LCR, the decarbonization of energy, heating, and transport through widespread adoption of renewable energies (REs), an apparently easily attainable task since technological capacity is available and policy instruments are already in place, is progressing much too slowly (European Commission, 2015). While progress has undoubtedly been made, there are good reasons to question whether Europe will be able to achieve its clean energy goals in time. The aim of the research described here is thus to explore the reasons behind the slow progress on implementing the clean energy transition by looking in detail at the *social process* that such a transition entails.

One explanation may be that existing policy instruments are not sufficient to secure implementation without the collaboration of key insiders, known as incumbents, who are resistant to change on the basis that it threatens the status quo (Hewitt et al., 2017). To achieve a transition to clean energy in the short window of opportunity available to avoid the most disastrous effects of climate change, the impetus must come from society as a whole, not just from policy-makers, engineers, or scientists. In fact, securing a transition to clean energy is not a technological, economic, or political problem, but a complex social process in which policy, economy, and technology are inextricably entangled. To deal with this complexity, a wide range of innovative and transformative approaches are required. Excluded stakeholders need to be brought in from the margins, politicians and energy companies need to operate more transparently, vested interests and traditional power structures need to be challenged, and markets need to be brought to serve the well-being of the

[40]Funded by the Seventh Framework Programme of the European Union (FP7), project coordinator Nick Winder. WP3: Making climate change policies work. OCT, University of Twente.

planet and its inhabitants, not just an increasingly small group of elite multinational institutions and individuals. None of this is easy, but the alternative is likely to be much worse.

The aim of the COMPLEX project was to explore how such an extraordinary social transformation might really be achieved. Citizens have become accustomed to the idea that the only way to make progress on this issue is through global conferences like COP 21 (21st Conference of the Parties to the Kyoto Protocol), which led to the Paris Agreement. Once these international arrangements are in place, the thinking seems to go, the rest (policies, directives, and instruments) will follow and the problem will begin to be solved. Unfortunately, our European experience suggests that such optimism is misplaced. Policies to reduce emissions and green the economy had been in place in Europe for many years prior to the signing of the Paris Agreement, but even very wealthy, technologically advanced countries like the Netherlands and Germany made little progress in the years leading up to Paris. Indeed, despite the great hopes for German energy transformation, the *Energiewende*, the closure of nuclear power stations in the wake of the Fukushima disaster has paradoxically led to increased emissions from burning cheap lignite or brown coal. The state-owned Swedish energy giant Vattenfall recently sold off some of its coal mines, cleaning its own image but doing nothing to reduce carbon emissions, since the mines were bought by a Czech company that plans to exploit them (Reuters, 2016). Clearly, the great discrepancy between the optimistic language of policy documents like the EU LCR (which, despite their technically binding nature, actually amount to little more than vague promises) and the reality on the ground in individual countries is likely to repay investigation.

Spain was chosen as our case study because of the great contrast between the very high level of development of RE systems achieved before 2011 (see, e.g., Ruíz Romero, Santos, & Gil, 2012) and the subsequent sudden reduction in RE development. This dramatic change came about because the newly elected conservative government of Mariano Rajoy, affected by the economic and financial crisis, turned its back on REs and even took the regressive step of halting new RE developments by cutting subsidies, removing feed-in tariffs, and disincentivizing battery storage for grid-connected household consumers.[41] As a result, Spain has rapidly changed from being a pioneer and leader in European RE development to becoming a pronounced retarder, and there is a risk that the country's key objectives set down in the LCR might not be met. In the light of this experience, many questions came to our minds. If such a thing could happen in Spain, we wondered, could it happen anywhere else? What particular conditions could have led to such a rapid shift in energy policy at precisely the moment when climate change mitigation efforts should be accelerating, not going into reverse? Most importantly, perhaps, we wanted to ask if a clean energy transition can really be achieved, either in Spain

---

[41]RD 1/2012, January 27, Law 24/2013, December 26, and RD 900/2015, October 9.

or anywhere else. And, if, so, what are the most important changes that need to be made to make this possible in the near future?

Our work under COMPLEX was thus focused on both an analysis of what had gone wrong and the search for ways to recover this positive trajectory and achieve a truly sustainable energy transformation. Our understanding of the clean energy transition as a social process implied that its success will be determined by the actions of key actors such as policy-makers, energy suppliers, and businesses. Our approach therefore took as its starting point the need to discover the identity of these actors and involve them as widely as possible in the co-construction of negotiated pathways to the future transition. To do this we employed a wide range of participatory approaches, including interviews, sociograms, and participatory spatial modeling, scenario-building and role-playing activities. Our aim was to construct a collaborative process that would lead, we hoped, to social learning, and allow us to identify opportunities and key objectives and challenges (Fig. 3.64) to move the clean energy transition forward.

To begin our work, we carried out a desk-based assessment in which we analyzed published, unpublished, and internet sources for the whole Spanish territory (17 autonomous regions and 2 autonomous cities). This initial survey revealed a great diversity of plans, policies, approaches, and implementation success stories that threatened to overwhelm the research team's capacity, so a filtering process was initiated to identify six key regions that were representative of this diversity for further study.

FIGURE 3.64

Detail of the working process in one of the workshops: participants are defining challenges and objectives for RE implementation.

3.3.3.2 Description of Techniques

3.3.3.2.1 Surveys and Interviews (Chapter 2, Section 2.1.2.1.2)
The team identified prospective key stakeholders both at national level and in each of the six identified regions, and contacted them by telephone. The key aims of this initial contact were as follows.

1. To engage main stakeholders with the COMPLEX project
2. To develop an understanding of the regional aspects of the implementation of RE policies.
3. To request digital cartographic data for regional Renewable Energy-related Landscape Features (RELF) and LUCC for GIS analysis and subsequent model development.

A "snowball" process (collective intelligence techniques, Chapter 2, Section 2.1.2.1.3) was initiated whereby stakeholders contacted facilitated access to additional more appropriate or knowledgeable stakeholders. The process culminated with a series of detailed semi-structured telephone interviews with these key stakeholders for each region and at national level. In all, 10 stakeholders were interviewed (Table 3.14).

Following the telephone interview phase, we chose one of these six regions, Navarre, for more detailed in-depth study through a series of workshops with key stakeholders. Navarre was chosen because of its long history of RE implementation, its diverse community of interlinked social actors around RE development, and because the stakeholders contacted were very interested in and receptive to the project.

3.3.3.2.2 Timeline (Chapter 2, Section 2.1.2.2.1)
The timeline technique was used to elicit information about key events and milestones since the beginning of RE implementation in Navarre. A blank paper chart was hung from the wall and a horizontal line drawn from left to right through the center of the paper. Above this line were the milestones and events which participants considered positive (favorable to the implementation of RE in Navarre); below the line were the milestones and events that participants considered to be negative (unfavorable to implementation of RE). Milestones and events, with their exact or approximate dates, were proposed out loud by participants as part of a group brainstorming session, and the opinions and reflections of all participants were recorded on the chart. Events were designated as positive or negative by the participants who described them. If there was disagreement about whether an event or milestone could be considered positive or negative, a red circle was drawn around the event to indicate this. The timeline activity produced a large amount of information and served as a starting point for interesting debates about RE in Navarre. Fig. 3.65 shows the process of developing the timeline, and Fig. 3.66 shows the final timeline digitized by the research team. Table 3.15 shows some of the most important observations that emerged from this activity.

Following the timeline activity, a *trendline* activity was carried out. The purpose of the trendline was to describe how people understand the processes that have

Table 3.14 Guide for Semi-structured Telephone Interviews With Regional Stakeholders

Guide for Semi-structured Telephone Interviews With Regional Stakeholders
Renewable Energies Tendencies and Current Situation
What type of REs is the region producing?
How has RE evolved in the region during the last decade?
Which factors are stimulating or inhibiting the use and implementation of REs in the region?
Is there any land-use change driven by the implementation of climate change mitigation policies?
Are there any conflicts about the use or implementation of REs? If affirmative, do these conflicts have any clear territorial implications? To which RE is the conflict related? What are the reasons for the conflict?
What is your opinion about RE development in the future?
Regional Policy Framework
Is there any regional energy plan in force? Has this plan defined any horizon or objective for REs? Have these objectives been achieved? Were these objectives achievable?
Is there any economic incentive linked to the RE promotion policies?
Is there any area where the development of RE installations is forbidden or limited?
Involved Stakeholders
Which social groups are affected by the RE development in your region?
Are there any organizations, centers, institutions, associations, or NGOs with relevance (with partial or total dedication) in tracing, support, research, or implementation of REs? Which?
Is there any local initiative about sustainability or climate change mitigation (e.g., sustainable villages)?
From your point of view, what role might your organization play in relation to RE development?
Cartography Availability
Is there spatial information about locations of RE infrastructure?
Are there regional spatial land-use datasets? If affirmative, how many time periods are there?

occurred over time, above all those that are related to development, such as climate change, changes in production, resource availability, income, etc. Identifying the change trend provides important information even if it cannot be quantified—for example, it may show how different groups can have different perceptions of the same changes (Geilfus, 2002). Participants were asked to draw, relative to the time-line, a trendline to show how the value and interest given to RE implementation by society as a whole had changed in Navarre. The activity produced an intense debate about trends and social perception of RE, but we ran out of time before the line could

FIGURE 3.65

From top to bottom: the workshop facilitator invites participants to contribute information to the timeline, which she writes on post-it notes and adds to the wallchart; facilitator writing down explanations and remarks; final timeline. The activity began with a single line on the wallchart and no other information. This activity served as an "ice-breaker." Participants were initially seated, but became more animated and involved as the timeline developed.

actually be drawn. This "failure" was no obstacle to the work because the discussion generated many interesting insights and a great deal of data.

The main benefit achieved from the timeline and trendline activities was to increase the research team's understanding and enable us to see the problem from the point of view of individuals and organizations with detailed knowledge of the situation. It can clearly be seen that the events described in these activities result from political decisions, from privatization of the region's public energy company to the government's decision to withdraw support for RE, resulting in loss of investment in RE due to legal uncertainty. At the same time the region's unusual tax status, in terms of its fiscal autonomy, explained to some extent the high level of RE development in the region, and may offer an opportunity for RE development to be restarted. Although this information could have been obtained in other ways, the process would have been much more laborious.

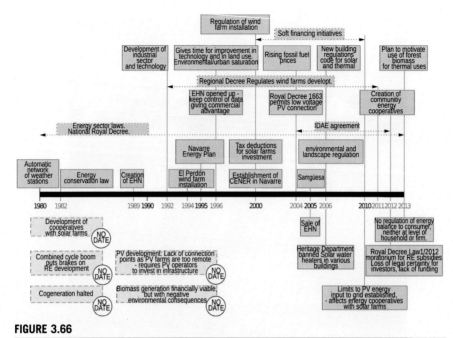

FIGURE 3.66

The final timeline for RE development in Navarre.

3.3.3.2.3 Sociogram (Chapter 2, 2.1.2.2.2)
Sociograms were developed in three phases.

3.3.3.2.3.1 Preliminary Sociograms
The team made several preliminary sociograms based on the literature review as part of the initial process of identifying the main social actors (groups, institutions, or individuals) involved in RE implementation in Spain. Beyond this simple objective, we hoped to identify the social relations that exist between them and highlight flows of information, relationships of trust, mistrust, or uncertainty, or any type of link between the actors that could be of interest in understanding the RE development process. We drew one sociogram for the national level and six at regional level, one for each of the key regions selected for more detailed study. Actors were represented using simple circles. Four larger circles were drawn in each corner of the paper representing the key sectors involved in RE development, business, education/research, civil society, and government/administration, and the actors were located within these groups. At this point we still had no understanding of the social relations existing between the various groups, institutions, or individuals identified, so no overlaps were shown between the sectors.

3.3.3.2.3.2 Intermediate Sociograms—Adding Information Provided by Interviewees and Analyzing the Sociograms
Following the semi-structured interviews, we modified the sociograms, adding in new stakeholders and moving the sectors to show overlapping areas of competence

Table 3.15 Workshop Participants' Observations Recorded From the Timeline Activity

- The regional government representative highlighted the importance of legislation and subsidies for RE development in Navarre.
- Decreto Foral 1996 (regional legislation) is emphasized at regional level as "the starting point for an avalanche of RE projects" in Navarre.
- Complementary tax benefits are highlighted in the regional context due to fiscal autonomy in this region, which allowed establishing deductions for personal income tax and corporate income tax linked to RE from the year 2000.
- Before 1994 the role of the Navarre Hydroelectric Energy semipublic company is noted. 1996, the year in which the energy sector was liberalized, was added as a key date associated with the tariff deficit.[42] The sale of the company in 2005 is noted as a negative factor for RE development in Navarre.
- Some reflections are made about solar farm cooperatives. They are considered both positive and negative (positive for emissions reduction but negative on account of their landscape impact). That is why this item is marked in red.
- The current loss of legal certainty is noted as a cause of lack of financing for projects.
- The regional government representative also highlighted the regional Forest Biomass Plan as an opportunity for the sector. Nevertheless, representatives of civil society organizations[43] note that although feasible, the biomass initiatives developed are environmentally unsustainable.
- One of the events marked in red is the increase in the price of fossil fuels in 2004. This was considered to be a driver for development of RE and related technologies, but at the same time participants observed that the impact on the population has been negative.

or interest (Fig. 3.67). Once the interviews had been completed the seven intermediate sociograms were analyzed, taking into account the following assumptions and general considerations.

1. Overlaps. The existence of overlaps which represent spaces of common work between different spheres of action implies a more participatory development process in which communication and information exchange between each sphere are stronger. Two of these overlaps seem to be of special importance: the link between business and administration, which is important for investor confidence, and the link between business and civil society, which ensures widespread diffusion of the RE implementation process throughout all sectors of society, not just among the elite. Strong links between organizations as a result of these overlaps may imply a higher level of resilience.

---

[42]The tariff deficit (*deficit tarifario*) is a source of much controversy. It refers to the difference between the price the government allows the electricity companies to charge the consumer and the price that electricity companies claim they need to charge to be profitable. This exists as a government debt to the electricity companies and appears as an asset on companies' balance sheets. Opinion is divided as to why this debt is necessary, whether it can be renegotiated, or whether it really benefits the consumer.

[43]E.g., environmental groups like Sustrai Erakuntza or Ecologistas en Acción.

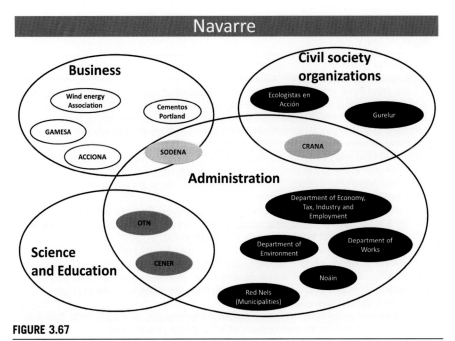

FIGURE 3.67

Intermediate sociogram for Navarre region.

2. The role of the regional administration. The regional government is important in the RE implementation process. It can act simply as another stakeholder, or it may choose to play a leading role. When it chooses the second option, it is linked to the other spheres of action.

3. The organization level of the business sector. As provider of the key financial resources, the business sector also has the potential to play a leading role. If this sector is highly organized around RE, this is likely to be positive for the further development of RE in the region.

4. Isolation of some spheres of action. The involvement of all spheres of action in the RE implementation process would contribute to a more complex and dynamic social network around RE. This is likely to increase the chance that the whole stakeholder community would resist the imposed destruction of the system by a single actor, and instead, when the system suffers a shock, try to generate different options for the future.

Analysis was carried out for all seven of the intermediate sociograms; only the example for Navarre is discussed below.[44]

---

[44]Complete results of the work can be found in Alonso et al. (2016).

The Navarre sociogram (Fig. 3.67) stands out in terms of the high level of participation of the regional public sector in the spheres of action, especially the existence of organizations that link the administration with all the other spheres. This is unique among the six regions studied in this research. Since there is no single workspace that is common to all spheres of action, this factor is clearly highly significant to RE development in Navarre. The business sphere is robust and well organized. Several very large businesses, such as Acciona, Gamesa, and Cementos Portland, are present in Navarre, which is likely to be influential in determining future RE implementation.

3.3.3.2.3.3 Participatory Sociogram

The final sociogram activity was carried out in the first Navarre workshop, with the aim of validating and further developing the information we had already collected from the desk-based research and telephone interviews. This was important because, up to that point, although interviewees had provided information that we had tried to represent on the sociograms, no stakeholder had actually seen the sociograms or intervened directly in their development. Prior to the workshop, our knowledge of the stakeholder community remained limited. The activity had four key aims.

- To enrich our own and other participants' understanding of the community of actors around RE by further developing our initial impressions recorded on the intermediate sociogram. We were particularly interested in learning about new actors we had not been able to discover.
- To identify *action sets* or interconnected groups of actors with a capacity to drive change.
- To describe the social linkages between the actors, drawing the relationships on to the diagram using lines and arrows.
- To understand how our participants viewed their own role within the community of actors involved in RE development.

Before the workshop took place, organizations that had confirmed their attendance were located within one of the four basic groups already identified on the intermediate sociogram. This sociogram was presented at the workshop, with participants identified on post-it notes. Our four provisional categories were business entities involved in implementation of RE, social and environmental organizations, scientific or educational organizations, and governmental or public organizations. These action sets clearly had overlapping "common zones."

Participants were asked to revise the sociogram and see whether the entities had been appropriately grouped. If they were wrongly placed, they were invited to locate them in the correct place by moving the post-it notes. Newer participants, who had not been included in our pilot study because they had joined the snowball later, were asked to draw the relationships between their own organization and other entities or organizations present at the workshops using different types of lines, as shown in Fig. 3.68 (information exchange, occasional specific collaboration, permanent collaboration, and disagreement/difference of opinion). Following this activity all participants were asked to identify on the wallchart, in black ink, any regional-level

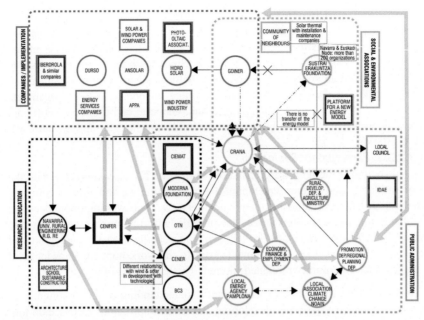

FIGURE 3.68

Final participatory sociogram, Complex project.

FIGURE 3.69

Representatives of the different action groups working on the flipchart, defining existing linkages between stakeholders.

stakeholders they felt to be key for the sector, but that were not identified on the sociogram, together with their relationship lines (Fig. 3.69). They were then asked to do the same, in red ink, for national-level stakeholders. Participants were encouraged to add comments to the sociograms where necessary.

3.3.3.2.3.3 Results

Participants added a large number of new actors to each of the broad groupings (spheres of action) and drew connections between them. Many strong connections were identified both within the government and public sphere and between this sphere and the other groups, especially research and education, where particularly effective links were identified with the National Renewable Energy Centre (CENER), the Moderna Foundation, and Navarre University. CENER is a publicly funded research and development organization specialized in REs, offering services to both public and private clients, while the Moderna Foundation is a not-for-profit organization, also with strong public financial support, specifically dedicated to promoting private—public sector cooperation to implement "objectives defined by society" (FM, 2016). Perhaps unsurprisingly in this context, strong links were identified between the government and public sphere and the RE business organizations' sphere.

During the previous phase researchers had tried to engage the private sector. Five energy company representatives were invited to participate: all expressed polite interest, but none provided an interview or attended the workshop. Their absence was quickly detected by the participants, who highlighted the need to include some of the large RE and engineering companies operating in the region, specifically Gamesa, Acciona, and MTorres (all of which have been heavily involved in large-scale wind installations), as well as Iberdrola, a major national energy supplier. Participants also suggested that the business sector should be divided into three subgroups to differentiate RE installation companies from infrastructure engineering companies and energy suppliers. In Fig. 3.68 (top left) one can see that workshop participants could identify relevant private sector organizations but were unable to establish links between them. This is presumably not because no links exist, but rather because we were unable to involve any knowledgeable insider from the private sector in our process.

3.3.3.2.4 Participatory Modeling (Chapter 2, Section 2.1.2.3.1)

One of the central objectives of this project was the development of a spatial model of RE development in the territory to understand how widespread implementation of RE would affect future land-use trajectories. As in the previous participatory modeling cases discussed in this theme, we approached this by developing a set of scenarios to reflect a range of plausible hypothetical outcomes by the year 2050 (the final date by which the ambitious objectives set down under the EU Low-Carbon Roadmap are expected to be achieved), based on the information we had gathered through the participatory process. The spatial model itself used the same underlying conceptual framework as the Metronamica model, described earlier, but unlike the previous projects we programmed the model ourselves using the free and open-source R environment. The model we developed was known as the actor, policy, and land-use simulator, or APoLUS (Hewitt et al., 2015).

Following the activities in the first participatory workshop, we moved from scoping, stakeholder identification, and problem framing toward building the spatial model. At this point work was begun on software development for the APoLUS model in R software in parallel with the second stage of stakeholder elicitation through the second workshop, denominated parameterization (WS2), which was held at the Museum of Environmental Education in Pamplona in June 2014. This workshop addressed four objectives: finding the appropriate land-use categories for spatial modeling of implementation; understanding multifunctionality related to land-use change and installations; geographical location of the key areas of development in Navarre; and refining and extending modelers' understanding of actor behavior and climate change mitigation relating to implementation through role playing of specific actors. At this workshop, the following activities were carried out.

1. Activity 1. Participatory reclassification of land-use categories and association with key RE types. This activity was undertaken in two parts, to address two objectives.
   a. Activity 1a: To decide which land-use categories are involved in the process of inclusion of RE infrastructure in Navarre (land-use classification exercise)
   b. Activity 1b: To decide in which cases one land use is substituted by another, and in which cases the energy-related use is added to the previous use (land-use compatibility with RE types)
2. Activity 2. Participatory cartography for the main land-use categories in Navarre.
3. Activity 3. Role-playing game related to land change dynamics and the principal drivers of change in RE development.

These activities are described below.

Activity 1a: Land-Use Classification Exercise

Before starting this exercise, researchers carefully explained the aim of the activity: to arrive at a land-use grouping that reflected the territorial realities of Navarre at a level of detail sufficient to understand how RE installations affect different land-use types. For example, to understand whether REs are preferentially installed on less valuable or productive land, it is necessary to group land-use categories in such a way that some general distinction between crop types is possible, e.g., irrigated crops and traditional crops, rather than "agriculture."

Participants were divided into two groups, with a third group formed by two members of the research team. Each group was given a printed dossier containing 11 basic land-use categories (see Fig. 3.3): herbaceous non-irrigated crops; herbaceous irrigated crops (Fig. 3.70); woody non-irrigated crops; woody irrigated crops; Mediterranean shrub; pasture/grassland; woodland; unproductive land and bedrock outcrops; urban land; infrastructures; and water. Beside each category there were

Cultivos herbáceos de regadío
Se incluyen cultivos herbáceos como maíz, trigo, cebada, arroz, hortícolas como el pimiento o el tomate, espárragos en regadío

¿Crees que deberíamos desagrupar esta categoría?

¿Qué criterio deberíamos utilizar?

FIGURE 3.70

Land-use classification dossier, herbaceous irrigated crops.

two questions for participants to consider carefully in their groups and agree a response.

> Q1: Do you think that this category ought to be separated out into other, more detailed categories?
> Q2: If yes, what criteria should be used (e.g., crop type, productivity, something else)?

After around 20 min of internal discussion, each group chose a spokesperson to explain the decisions their group had made to everyone else. The categorization decisions made by each group were written down on a wallchart.

Activity 1b: Land-Use Compatibility With Renewable Energy Types

Once a list of categories had been agreed by debate and consensus between all the groups, participants were asked to reflect on the relationship between these land-use categories and three main types of RE infrastructure (wind, solar, and biomass). The relationship between land use and RE installations was considered to comprise two basic types: combined use, in which the new energy use is incorporated into the existing use, for example the installation of solar panels on rooftops in urban residential areas; and substitution of the existing land use by the new energy-related use, for example a solar array installed on agricultural land where the previous use for food production is entirely replaced by the new use for energy production.

Three wallcharts were prepared, one for each energy type—wind, solar, or biomass. Each chart had four columns: on the left the agreed list of categories from the previous part of this activity, and to the right three columns to reflect compatibility, substitution, or either of the two (if an energy use could either combine with or substitute a particular land use, depending on the case). Each group completed each of the three charts using a different colored pen (green or red). Groups were asked to mark the relationships currently found in Navarre with an x, and those which are not currently found but which it would be desirable to see with a "heart" symbol. One group decided to include a "prohibited" symbol to reflect relationships which ought to be avoided in future even if they are not currently known to occur.

In the biomass wallchart, one group decided to distinguish between the land use which provided the natural resource and the land use where the biomass would be processed.

Activity 2. Participatory Cartography for the Main Land-Use Categories in Navarre

The aim of this activity was to locate, through a participatory cartography exercise, those areas of Navarre in which such substitutions by RE or combinations with RE are likely to occur on existing land-use maps. Seven A0 land-use maps were provided, each showing one of the agricultural *comarcas* (counties) of Navarre (historic areas of production relevant to agriculture, but with no modern administrative role) (Fig. 3.71). Participants were given three blocks of differently colored post-it notes, with each color representing a different RE type (yellow for solar energy, orange for

FIGURE 3.71

Participants working on the participatory map.

FIGURE 3.72

Detail of the map development process.

biomass, and green for wind energy). In the same groups as previously, participants circulated freely among the maps discussing the issues and localizing the post-it notes in areas where they knew RE infrastructure to be located (Fig. 3.72). On each note participants wrote information about the relationship (combination, substitution, or either) between the RE type and land use at that location. Participants also included post-it notes where they considered that RE infrastructures ought to be developed, indicating this with a heart symbol. By combining the seven county maps containing this participant information, an initial participatory map of land use and RE was created for Navarre. This map did not aim to be exact or definitive, but was rather a preliminary attempt to collect and express in a spatial format stakeholder knowledge and perceptions about the relationship between REs and territory in the region of study.

Activity 3. Role-Playing Game Related to Land Change Dynamics and the Principal Drivers of Change in Renewable Energy Development

The aim of this activity was to reproduce simply some of the possible ways in which the highly complex interaction process of RE implementation in Navarre might proceed, to understand better the ideas, interests, arguments, and ultimately the motivations, resources, and power developed by the different actors involved in decision-making over land use.

The role play attempted to replicate, in a simple way, the process of negotiation between key stakeholders involved in RE implementation in the territory. Two "situations" related to this implementation process had emerged as key challenges in the previous workshop and were used to define the context in which the role play would be set: the perceived "saturation" of the landscape by large-scale RE installations (e.g., multi-hectare wind farms), and the occupation of rural land not classified for development by RE installations. The situations were chosen because of their clear potential for opposing viewpoints and interesting negotiation possibilities.

Workshop participants were each allocated a role, usually a different or conflicting role to their normal professional area of action, and given a picture card representing this role. They were then encouraged to play their role as realistically as possible, according to their own understanding, through any type of interaction with other participants that they perceived to be appropriate. The different roles established after the first workshop as a result of the participatory sociogram exercise were regional government; local government; landowners; environmentalist groups; large electricity companies; the national electricity grid operator; wind, solar, and biomass sector SMEs; consultants; local community groups; civil society; energy production and distribution cooperatives; scientists and researchers; and the media. However, as there were not as many participants as roles, some of these roles were played jointly: energy production and distribution cooperatives; landowners; wind, solar, and biomass SMEs; environmentalist groups + civil society + local community groups; large electricity companies; and regional/local government (Fig. 3.73). Note that both "regional/local government" and "large energy companies" were played by researchers, on the basis that neither researcher had experience in these sectors, and playing this role was therefore a more creative learning experience that might have been the case if participants had played their own roles.

Following the activity, each participant was required to explain publicly his/her role and the position he/she had adopted to represent the role as faithfully as

Grandes Empresas Energéticas

Red Eléctrica Española

FIGURE 3.73

Two of the roles stakeholders were asked to play: large energy companies (left) and the national electricity grid operator (right).

Table 3.16 Questionnaire Completed by Players at the End of the Role-Playing Game

Name of player:	**POWER** (how much power did you have and how have you used it?):
Role in the game:	**RESOURCES** (what resources did you have and how have you used them?):
Challenges to deal with (tick)	**MOTIVATIONS** (Which motivations drove the negotiation process?) tick one or several boxes and explain:
☐ Occupation of rural land not classified for development by RE installations.	☐ - ECONOMIC
☐ "Saturation" of the landscape by large scale RE installations (e.g. multi-hectare wind farms)	☐ - POLITICAL
	☐ - SOCIAL
Original position in respect of the challenge (before the beginning of the game):	☐ - CULTURAL
Have you achieved any agreement with other stakeholders to deal with the specific challenge? (yes/no)	☐ - ENVIRONMENTAL
With whom (which roles) have you negotiated?	☐ - OTHERS
Did your original position change as the game progressed? How?	
Describe the negotiation process	

possible. Participants also completed a questionnaire (see Table 3.16) explaining the role they had played, with whom they had negotiated and why, and what their motivations had been in responding to the situation defined.

The role-playing game took the form of a simulated meeting between key actors. All participants played their roles enthusiastically for around 60 min. Each participant explained to the whole group her/his role, and a debate subsequently developed in which different visions and positions relating to the two situations emerged. During this time participants also separated into smaller groups and simulated private meetings to seek agreement. The activity finished with a conflict resolution exercise in which participants explained the decisions they had taken during the negotiations, the motivations for these decisions, and the possible pathways available to them (power and resources) to move forward to consensus (depending on their motivations).

Some of the most interesting results are presented in Table 3.17a–f. As a conclusion we can say that none of the situations which emerged was unpredictable, but the role play did suggest that consensus could be reached on several issues among different and diverse stakeholders.

Activities 1 and 2 were extremely useful for eliciting stakeholder knowledge on the relationship between land use in Navarre, generating essential data for parameterization of the land-use block of the APoLUS model. The role-playing game (Activity 3) generated a substantial amount of information about the positions that might be adopted by the various actors, which was very useful for simulating different implementation situations (relative to actor motivation, power, and resources) in APoLUS.

Table 3.17a Position Adopted by the Energy Production and Distribution Cooperatives Player A

Player A: Energy Production and Distribution Cooperatives	
POSITION ADOPTED To promote 80% of energy consumption under a distributed system	**POWER: LOW** Very low power, as this area of activity is quite isolated because of new legislation
STAKEHOLDERS TO WORK WITH Landowners, local and regional governments, SMEs, environmentalist groups	**RESOURCES: MEDIUM** Cooperatives and RE farms; the problem is the lack of physical connection between the points of production and points of distribution
NEGOTIATION PROCESS Landowners have the land, the SMEs build the farms, the cooperatives distribute the energy, the local and regional governments favor these new instruments for energy business and keep the big energy companies out of energy distribution at this small scale	**MOTIVATIONS: HIGH** Economic: support energy self-consumption Political: looking toward a change of model Social: it is a new opportunity to work together as a community for a social economy Cultural: management of local resources (sun, wind, biomass) as services for the community 5th. Environmental. Searching for a sustainable model

Table 3.17b Position Adopted by Landowner Player B

Player B: Landowner	
POSITION ADOPTED Increase the value of my land	**POWER: MEDIUM** Land
STAKEHOLDERS TO WORK WITH Small cooperatives for energy distribution, local and regional governments, SMEs, environmentalist groups	**RESOURCES: MEDIUM–LOW** Possibility of negotiating with big energy companies and SMEs
NEGOTIATION PROCESS A common agreement between SMEs, cooperatives for energy distribution, local and regional governments, and landowners This strategy will may not achieve highest market value for my land but I can try to offset this in other ways such as a certification of ecological production I have supported this agreement because I feel defeated by the power of the big companies, although I am still doubtful about the real possibility of this type of energy market becoming a reality	**MOTIVATIONS: HIGH** Economic: support energy self-consumption Social and cultural: to be opposed to RE is unfashionable at the moment; also I am very attached to my land and would like to keep farming, so I would prefer to invest in some activity which would allow me to enhance multifunctionality on my land, and RE is one possibility

Table 3.17c Position Adopted by the Wind, Solar, and Biomass SMEs Player C

Player C: Wind, Solar, and Biomass SMEs	
POSITION ADOPTED Promote small-scale local REs and try to extend their implementation to urban and industrial areas	**POWER: MEDIUM** Create employment
STAKEHOLDERS TO WORK WITH Small cooperatives for distribution, local and regional governments, landowners, environmentalist groups	**RESOURCES: MEDIUM** SMEs in Spain are the main source of employment (60% of the workforce are employed by SMEs)
NEGOTIATION PROCESS "Together we are stronger" is the motto for this process to stand up to the big energy companies; the support of the government is essential in this process as we want to implement RE but with better environmental sustainability criteria than in previous development phases	**MOTIVATIONS: MEDIUM** Economic, environmental, and social

Table 3.17d Position Adopted by the Environmentalist Groups + Civil Society + Local Community Groups Player D

Player D: Environmentalist Groups + Civil Society + Local Community Groups	
POSITION ADOPTED In favor of microscale production and consumption	**POWER: HIGH** Social lobby
STAKEHOLDERS TO WORK WITH Energy production and distribution cooperatives, SMEs, landowners, local and regional governments, agrarian trade unions, social platforms	**RESOURCES: HIGH** Public campaigns Environmental awareness
NEGOTIATION PROCESS A pressure campaign to get together with the other stakeholders in favor of RE to promote more affordable and cleaner energy, create better value for farmland and production, and bring the local/regional government onside to develop a "green label" to tackle climate change	**MOTIVATIONS: HIGH** Political, social, environmental, and economic A change of the model of distribution to be more equal and favor strategies to tackle climate change

Table 3.17e Position Adopted by Large Energy Companies Player E

Player E: Large Energy Companies	
ROLE and POSITION The implementation of REs and how these affect land uses does not bother me at all I do not want low energy prices or competition from energy cooperatives as this affects my profits, so at the moment I am not interested in further RE implementation anywhere in Spain, although I am certainly interested in "cleaning my public face"	**POWER: HIGH** Lobby influence Control of energy distribution
STAKEHOLDERS TO WORK WITH Public administration at all levels	**RESOURCES: HIGH** Money for investments Big production and distribution of energy
NEGOTIATION PROCESS This chaotic situation can only favor us, the big companies, as we will keep extending our business outside Spain while we try to work on publicity campaigns and support politicians through efficiency energy campaigns	**MOTIVATIONS: LOW** Political lobby: to keep politicians happy Social: an image-cleaning campaign

Table 3.17f Position Adopted by Regional/Local Governments Player F

Player F: Regional/Local Governments	
ROLE and POSITION Neither in favor of nor against RE, provided that a constant electricity supply is maintained; though nominally concerned to appear "green" provided this does not require additional resources, in reality I am not that interested in RE	**POWER: HIGH** High level of decision-making process "All stakeholders want to negotiate with me"
STAKEHOLDERS TO WORK WITH Big business is good business because it is good for job creation Disinclined to support small or local-based initiatives which promote a change of model Growing interest in taking into account coalitions of SMEs, cooperatives, environmentalist groups and so on (for political reasons)	**RESOURCES: HIGH** Budget Capacity to regulate and generate instruments
NEGOTIATION PROCESS Agreement with big energy company to favor energy efficiency (instead of RE developments) in rural areas Talks about future agreements with a coalition to promote economic growth by implementing RE in rural areas using EU rural development subsidies	**MOTIVATIONS: MEDIUM** Economic and political Needs new ideas to get the economy going Concerned about the regional budget and the economic crisis

Table 3.18 Narratives for Each of the Three Scenarios

Scenario 1: The Big Fish Run the Game

REs have become cheaper to produce and before 2012 were increasingly attractive to households and businesses. At the same time energy demand has fallen due to the economic crisis that Spain has been suffering since 2007. As a result, shareholders of the large energy companies have become concerned by falling profits and lobbied for removal of incentives (feed-in tariffs and tax breaks) to develop RE. A government has been elected with strong ties to the energy industry, less public money to spend, and a mandate to reduce the deficit from Brussels. Three measures were introduced by parliament between 2012 and 2014.

1. Paralysis of all subsidies to RE, sending strong signals to investors that the government will not support further development of the sector.
2. Introduction of a fee per KWh charge to be paid by any consumer wishing to connect to the electricity grid. This makes energy self-sufficiency unattractive to households and businesses.
3. Prohibition of the Tesla Powerwall battery in all installations connected to the electricity network. This means that consumers cannot use the battery to store energy they produce, and further limits the attractiveness of energy self-sufficiency to consumers.

No substantial change is expected before 2050 under this narrative. However, the government of Navarre has indicated that it may be prepared to use its regional autonomy over tax collection to divert funds toward promotion of RE at some point in the future.

Scenario 2: Regional Green Consensus

The economy improves but citizens are fed up with the government anyway and vote for change. The new government wants to promote RE development and restores previous favorable policies and financial incentives. The Navarre region also has a new party in control, with strong credentials in green energy and a commitment to fight climate change. RE in Navarre until now has mostly been developed in southern and central areas of the region, with little or no development in Pyreneen or Atlantic areas or around the regional capital of Pamplona<>Iruña; see map of municipalities and spatial planning zones (POTs). However, the new regional government modifies planning priorities to allow some (limited) development in other POTs where suitability for RE is sufficiently high. Priority is given to zones with a currently low level of installed RE, provided that the developments are undertaken in accordance with environmental legislation and environmental impacts are minimized. This scenario seeks to address the 2050 climate targets seriously while at the same time trying to avoid impacts to the natural environment through strong spatial planning at the regional level. Despite good will on all sides, there is very high potential for conflict over RE development in natural protected areas and their vicinity.

Scenario 3. Local Action for a World in Crisis

The economy worsens dramatically and climate change impacts prove more severe than expected, leading to a change of national government. The new government is broadly favorable to RE and tries to promote its development, but the economic situation means that subsidies are not restored. The parlous state of the public finances means that the national electricity grid is privatized, and immediately purchased by large energy companies seeking to entrench their control even further. The worsening economy, heavy cuts to social security (e.g., pensions), and high energy prices mean that a significant proportion of citizens in Navarre opt for energy self-sufficiency. Social movements take control of many town councils and establish small independent electricity networks. By the

Continued

Table 3.18 Narratives for Each of the Three Scenarios—cont'd

time the large energy companies realize that they are pricing themselves out of the market, it is too late to restore consumer confidence and some of them go bust or are split up into smaller units. By 2050 energy production and distribution in Spain have become highly decentralized, with considerable variation even between neighboring municipalities. This narrative is characterized by the formation of *ad hoc* alliances between municipalities sharing particular characteristics (e.g., rurality, poverty, proximity to the capital city, etc.) to take control of their own energy generation and distribution, and a high level of conflict between actors like big energy, which fight ever more fiercely to maintain their monopoly, and cooperatives, town councils, and citizen groups.

Scenario Development

As the participatory process was already quite extensive, the scenarios were developed by the research team on the basis of information obtained in the research process.

The situation in Spain at the time of the project, in the context of the detailed analysis provided by the stakeholders during telephone interviews and the first workshop, led us quite naturally to "Scenario 1: the big fish run the game" (Table 3.18). In WS1 the wide range of challenges identified by the stakeholders and the solutions they proposed allowed a second scenario narrative to emerge. This was initially baptized "yes we can (together we are stronger)" and later became "Scenario 3: local action for a world in crisis" (Table 3.18). The detailed information about deployment in the different regions of the study area allowed a third scenario to be developed to reflect the widespread concern about landscape planning. This became Scenario 2: regional green consensus (Table 3.18). At this point it was now possible to build and calibrate the APoLUS model for the Navarre case study region.

3.3.3.3 Conclusions and Lessons Learnt

The participatory modeling work carried out in Spain gave the team a much richer and deeper understanding of the RE implementation process than could have been obtained through non-participatory research activities. The detailed recommendations that have been formulated for the policy briefings are a direct result of this (Hewitt & Hernández Jiménez, 2016). We have a large amount of information on actor behavior around the implementation process, public policy with respect to REs at various levels, and land use and RE installations. All of this means that the model is well calibrated against real-world situations and the three scenarios should be highly plausible. We obtained such a large quantity of information that we were able to write a full-length research paper from the results of the telephone interviews and scoping statement alone (Alonso et al., 2016). We are still working on research outputs from the workshops and a publication on the model itself.

The poor attendance (five stakeholders) at the second workshop was principally a reflection of the date, just 1 week before the most important regional public holiday, the San Fermines. Stakeholders were impossibly pressured for time the week before this holiday, which they described as "the end of the world." This highlights the importance of local knowledge in any participatory process and the risks inherent in participatory research. Sometimes it is impossible to reconcile all the different requirements of the stakeholder community.

Conclusions: reflections and future prospects

4

4.1 INTRODUCTION

In this book we present and discuss a range of practical working strategies and tools aimed at changing current land and natural resource management policy and practice. To achieve this we have focused on the empowerment of stakeholders through participatory processes. The objective is to facilitate a social transformation characterized by three goals.

1. Greater awareness of the value of the territory and the multiple threats that it faces from the current "growth without limits" model of liberal consumer capitalism.

2. A paradigm shift in the way that urban and rural areas relate to each other. This involves reinforcing the interdependence of cities on their neighboring regions so as to favor proximity, regional identity, food sovereignty, food security, and sustainability over global strategies for increasing return on capital. This involves putting profit to communities before profit to external owners and investors, and fully integrating so-called "externalities" like environmental pollution into the cost of products.

3. Greater involvement of civil society in urban and regional planning, up to complete strategic control of this area in some cases, to end the exploitation of natural resources by external agencies and institutions in global markets without the full consent of local inhabitants. Organizations which extract for external benefit only, leaving behind devastated, transformed, or contaminated lands for others to deal with,[1] lose their social license to operate (Moffat & Zhang, 2014; Owen & Kemp, 2013). But in reality only rarely does this have genuine consequences. By transferring the power to decide on these kinds of developments to the community, social license becomes a requirement.

---

[1]This has been a key element in the fracking ban implemented in the state of Victoria (Australia), where it was observed that for every 18 jobs lost in the agricultural sector as a result of fracking, only 10 were generated in the gas industry. Furthermore, most of the benefits of this industry went outside the country (The Guardian, August 30, 2016 https://www.theguardian.com/environment/2016/aug/30/victoria-to-permanently-ban-fracking-and-coal-seam-gas-exploration).

The case studies presented in the preceding chapter pursue these three goals in diverse ways based on the requirements of each particular situation and the resources available at a given moment in time. Sometimes emphasis has been on the transfer of knowledge between social actors, feeding the system with the information they need to be able to find solutions and realistic alternatives to the problems in their territories. On other occasions we have worked on the design and delivery of participatory processes to facilitate the revitalization of rural areas at the local scale. And in other cases we have made use of more specific tools, like participatory modeling or mapping, as a means to drive forward innovative and consensual approaches to planning our environment.

The overall working strategy behind all the approaches presented in this book is necessarily adaptive—to the characteristics of each place, to the stakeholder community, and to the objectives defined under each project. Nonetheless, as we try to show, the broad theoretical and methodological framework introduced in Chapter 2 is a useful means of bringing together many apparently diverse initiatives into a single coherent process design. We do not aim to pronounce the last word on the subject, nor do we claim that the approaches we have applied are uniquely suitable. For this reason, in this final chapter we offer a brief reflection on the benefits and limitations of the approaches we have presented, framed as answers to questions that have occurred to us or to our colleagues during the process of writing this book.

4.2 WHY PARTICIPATORY ACTION RESEARCH?
4.2.1 ALTERNATIVES TO GROWTH WITHOUT LIMITS

Techniques and concepts from participatory action research (PAR) have been at the forefront of many of the cases presented in this book. PAR, and related methods like participatory rural appraisal, arose largely as a critical response to technocratic, externally led, and in many cases socially unjust approaches to development in the global South (Borda, 2001, pp. 27–37; Chambers, 1994b). Yet, as will have become clear from our case studies, rural development in Asia, South America, or Africa has not been our primary focus. The reason that PAR is also relevant in the rich world is because here we find similar problems; that is to say imposition from the top down of a growth-without-limits vision of territorial development without adequate consideration of possible alternatives. The alternatives, which have long existed (Schumacher, 2011 [1973]), have been given many names,[2] and have been approached from diverse perspectives, are all essentially variations on the same idea: living within the natural constraints of our Earth's system such that the Earth does not degrade over successive human generations. These alternatives

---

[2]Sustainability, *buen vivir*, living within planetary boundaries, etc.

are sometimes seen as idealistic or impracticable[3] because they are, by their nature, hard to accommodate in today's hyperliberalized consumer capitalism (Kemp & Martens, 2007). But the difficulty with limiting alternatives to solutions that do not imply any real change to the existing model is that the model is part, if not the root cause, of the problem. The notion, much in vogue today, that pro-growth, market-based approaches must be prioritized over all other possible approaches to governance predates widespread understanding about the commons, sustainability, or planetary boundaries, and leaves us ill-equipped to deal with the growing ecological crisis.[4] Thus the continued adherence of governments around the world to the principles of *laissez-faire* economics and the globally dominant liberal ideal of the state as a neutral facilitator of free markets mean that opportunities to make real progress on multiple sustainability issues, for example clean energy (Kuzemko, 2016), are being lost even as we reach, and surpass the point of no return for the planet. We need to be bold enough to consider options that look outside this narrow paradigm.

4.2.2 THE FAILURE OF INTERNATIONAL AGREEMENTS AND TOP-DOWN POLICY FRAMEWORKS

At the same time as we realize that we cannot rely on the ordinary workings of our governments to deliver us from disaster, we can also see that large international agreements, that most 20th century of answers to global problems, are not likely to produce a solution, at least not on a timescale adequate to ensure the well-being of our planet. The failures of Rio, Kyoto, Copenhagen, and most recently Paris[5] to shift humanity on to sustainable pathways suggest that while global environmental summits are a useful mobilization and dissemination tool, they will not on their own produce transformative change. Our social systems are locked in to pathways that prevent most of our largest institutions from acting for the common good.

It is clear, then, that we cannot depend on centralized sustainable planning initiatives mandated from the top down, either from supranational bodies like the European Union (EU) or the United Nations, or from our own governments. Policy instruments like the Habitats Directive (http://ec.europa.eu/environment/nature/legislation/habitatsdirective/index_en.htm), the European Landscape Convention

---

[3]Bjorn Lomborg, author of *The Skeptical Environmentalist*, is a good example of one such critic. See, e.g., Lomborg (2001).

[4]Even Hayek, one of the founders of modern liberal economic thought, does explicitly recognize that the principles of market competition cannot usefully be applied to what later became known as the commons: "Nor can certain harmful effects of deforestation, or of some methods of farming, or of the smoke and noise of factories, be confined to the owner of the property in question or to those who are willing to submit to the damage for an agreed compensation. In such instances we must find some substitute for the regulation by the price mechanism" (Hayek, 2014 [1944] p. 40).

[5]We do not wish to judge the outcome of the Paris Agreement in negative terms at a still comparatively early stage. But the absence of real incentives to cut emissions, in combination with the tremendous inertia in the system and in many countries the existence of a range of incentives to the contrary, means that success is by no means ensured.

(https://www.coe.int/en/web/landscape), and the EU Low Carbon Roadmap (https://ec.europa.eu/clima/policies/strategies/2050_en) provide welcome and necessary frameworks for action. But we should not confuse the existence of these kinds of environmental policy frameworks or conservation legislation with successful action. In one of our case studies, Doñana (see Chapter 3, Theme 3.3, Case 3.3.2), we found that a natural area which has enjoyed protected status since the 1960s under a battery of national and international laws, and is in receipt of numerous financial support measures designed to ensure its conservation, is becoming increasingly degraded. Doñana serves as a metaphor for our planet. Unfettered global trade binds us into pathways of ever-increasing unsustainable consumption, which goes hand in hand with ever-increasing damage to planetary life-support systems. However, it is unreasonable to expect institutions and power structures dependent for their survival on the continuance of this model to step in and solve the problem on their own.

4.2.3 BUILDING NETWORKS OF COLLECTIVE ACTION

Yet as individual citizens we cannot achieve these wider goals either. The problem is far too big and the system far too complex for individual actions like changing our brand of washing powder or choosing not to take our shopping home in a plastic bag to make much of a difference, and in any case the market on its own does not provide enough alternatives.[6] The path that remains to us is to build networks of collective action, beginning with local-scale initiatives and working outwards and upwards until we eventually transform the system around us. A wealth of successful grassroots initiatives reminds us that this is possible, from the transition towns movements to community energy initiatives and social mobilizations like the coalition of local residents of Torrelodones, Madrid, who began by protesting against an unwanted urban development and eventually formed a political party and won the municipal elections (Cueto & Muñoz, 2016). In this context, action research approaches, which strongly emphasize the role of local communities in collective problem solving, come to the fore. PAR, as we have seen in this book, brings to the table a wide range of practical tools for facilitating knowledge- sharing and building collective solutions. But the real strength of PAR in addressing these complex challenges lies not so much in its local-scale focus or its strong and well-developed methodological toolkit, but in its explicitly political dimension. PAR recognizes inequalities of power and works to address them by breaking down barriers and hierarchies and fighting to empower marginalized actors. This political dimension, though routinely forgotten in the name of dispassionate objectivity, is key to understanding how society interacts with the territory. Thus while the change needs to come from the bottom up, it cannot be successful unless it is accompanied by political changes at higher levels.

---

[6]The reasons why this is the case are discussed by Kemp, Schot, and Hoogma (1998).

Finally, PAR approaches are prominent in our work because of their flexibility. Participatory processes occur as a continuous sequence, or cycle, of action—reflection—action, in which working techniques are adapted as the project progresses. Once the problem has been identified and a general approach to the work has been decided, techniques are then chosen on the basis of a range of factors, e.g. the specific aims of the project, the characteristics and needs of the stakeholders involved, etc. As we have shown, particularly in Chapter 3, Theme 3.3, the process is quite organic, with simple techniques like flow diagrams gradually merging into more complex identification of problems, actions, and objectives, or techniques for diagnosis of a problem being turned on their head to evaluate our *approach* to the problem instead.

4.3 LIMITATIONS AND LEARNING EXPERIENCES

It is important, however, to recognize the limitations of what we are trying to do. Bringing about transformative change in land and resources planning is an ambitious goal, and often, wrapped up in an individual project or case, we may feel that we are very far from this goal. We do not wish to pretend that our approach has no limitations, or that all participatory processes are successful. We therefore offer the following general appreciation of some of the most important difficulties we have typically encountered.

4.3.1 WHO IS A STAKEHOLDER?

It is still common to find, as did Prell and colleagues in 2007, "resource management exercises [that] refer casually to stakeholders as if their existence and identity were self-evident" (Prell et al., 2007). But selecting and identifying the stakeholders is a key part of any participatory process. It is important to strive to maintain some degree of balance (Chapter 2: Section 2.1.1). A gathering of academics is a conference or a seminar, not a participatory process. A meeting of policy-makers behind closed doors may be an essential part of the management of a natural area, but it is not on its own a participatory process. Thus the credibility and validity of the participatory process are a direct result of decisions about who to include or not to include as stakeholders, and these decisions should be transparently declared and properly justified.

However, the struggle to facilitate the inclusion of stakeholders is a necessarily continuous process that is rarely accomplished perfectly, and is often in practice unsuccessful in its aim of maximizing inclusivity. Some processes, particularly if they are high-profile or perceived as successful, may suddenly find themselves overwhelmed by insiders (incumbents) keen to take some of the credit, which not infrequently drives away unacknowledged or marginalized groups. On the other hand, processes that only involve unacknowledged stakeholders may become echo chambers where everyone agrees with each other but no progress can be made because no one else has been invited. The question of stakeholder representativeness should not paralyze the project, but should certainly be a key element in the evaluation of the success of the process.

4.3.2 WHO KNOWS BEST?

As social researchers, we like to think about our stakeholders in terms of sectors and professional competences. Thus if we wish to achieve a good balance of opinion about, say, a proposed dam project, we might invite local residents who will be affected in some way by the scheme, an engineer with technical knowledge of the project, local farmers who may see some impact on lands they own, a river ecologist to advise about impacts on local aquatic ecosystems, a politician whose department is responsible for planning approval, etc. But if we are not very careful we can easily run into difficulties. For example, we may not see that the local residents are indifferent to the impact of the scheme because they feel it will create local jobs, the farmers may be delighted to sell unproductive land at greater than its open-market value under compulsory purchase, the ecologist may be employed by the scheme's developers and under instructions not to share "alarmist" results too early on, and the local politician may be the only one of eight local councilors in favor of the scheme. Thus an apparently representative selection of stakeholders may find themselves in unanimous agreement about an issue which has divided the community for decades, greatly undermining the chances of building true consensus. On the other hand, if nearly everyone present is sympathetic to environmental concerns but the community is a whole is deaf to the issue, no worthwhile progress may be made either.

   A related aspect is the question of who is really qualified to speak on an issue, and the unconscious bias that affects our decisions on these aspects. For example, is the opinion of an atmospheric chemist who has documented the effect of CO_2 on climate intrinsically more valid on that topic than that of a farmer who has not studied the issue and whose only information about it comes from a sensationalist news channel? Most people would agree that it is. However, if the same atmospheric scientist is invited to a workshop on agricultural land use, ostensibly for her knowledge on the carbon storage capacity of cropland, something she in fact knows very little about, does her information continue to be more valid than that of the same farmer, who cultivates and sells the crops under discussion? In fact, it is just as absurd to pretend that scientific information is always more valid than any other class of information as it is to pretend that it can never be more valid than any other class of information, since the issue depends entirely on the context. Human nature makes the problem even more difficult. On the one hand we frequently find people with great knowledge and expertise about key aspects who lack self-confidence and prefer not to volunteer this information. But the opposite case is also very common, where stakeholders are consulted over an issue about which they are largely ignorant, but do their best to volunteer an opinion anyway. This is like asking for directions from a stranger to the neighborhood who tries their best to help but sends you miles in the wrong direction!

4.3.3 TIME AND RESOURCES

Participatory processes can be very time consuming. Building a participatory model, for example, will typically take much longer than building a non-participatory one, and each stage of engagement with stakeholders will generate volumes of data that needs to be sifted through, systematized, and synthesized.[7] The disadvantage of this kind of rigorous, data-intensive approach is that long periods of time may elapse between cycles of participatory research, and stakeholders may lose interest, leave their organizations, or be promoted to new roles which reduce their opportunity to participate in the project. This can be compensated for to some extent by sharing information and providing feedback little by little as the project develops, rather than waiting for key milestones to share more substantial updates.

It is also common to allocate insufficient resources to participatory research, and the myth still persists that a participatory process can be added on to an almost finished project once all the other work has been completed. Even if the goal is evaluation, the work is better undertaken as a part of ongoing activities. Reporting can of course be left until the end, but not stakeholder identification or problem-framing activities. Clearly, if the aim is for stakeholders to provide meaningful input as part of the process of knowledge development, they need to be involved in the process as early as possible.

4.3.4 RECORDING AND DOCUMENTATION

We have occasionally found ourselves invited to participate in workshops at which no information is actually being collected. In seminars or conferences this is usual, but in workshops with structured knowledge-sharing activities it does not make sense not to collect information; at best it is a wasted opportunity, and at worst it completely invalidates the purpose of the activities. To be able to incorporate stakeholder information in research, information needs to be properly collected and attention given to the process of collecting it. Even if the only purpose of the workshop is dissemination of information, it is a waste of a good opportunity not to include at least one structured activity aimed at collecting information from participants. Usually, while note-taking, video-recording, and photography are essential for providing supporting documentation, the best way to get detailed information is from the participants themselves on *pro-formas*, post-it notes, and directly on wall-charts. Often information is anonymized, so it may be useful to give each piece of written information a unique identifier.

---

[7]It is usually a good idea to write a rapid report for the benefit of participants and to act as an *aide memoire* in advance of subsequent, more detailed analysis of the results. We have often come to rely on these kinds of reports, and we find in practice that even successful activities and processes tend to get forgotten if they are not rapidly recorded in this way.

4.3.5 THE PROBLEM OF INTEGRATION

Combining discursive, or participatory, information with quantitative data can be very challenging. Many environmental models, for example land-use models Fig. 2.3 typically combine both types of information, and therefore offer a good opportunity to bridge hard and soft science domains.

Working out ways to do this successfully is one of the most exciting and rewarding aspects of participatory modeling, but it is important not to be naïve about the difficulties that it can pose. There are also different types of integration. Technical integration, for example, in the sense of adding a parameter that has been estimated by stakeholders to a set of parameters obtained by numerical measurement, is often quite a trivial task, but genuine knowledge integration, such that different stakeholder communities actually assimilate one another's knowledge and approaches, is likely to be much more difficult. Not everyone understands the wider goals of participation. Senior scientists, for example, may struggle to move outside their comfort zones as respected experts and trusted providers of information to society. This can limit their ability to think critically and can be a barrier to genuine participation. This issue is frequently found with professional specialists of all kinds, for example government planners, who may be loath to admit that a plan or policy is not working and blame all failure on dissidents. There is a good deal of skill in learning how to manage these different perspectives, and there are many occasions on which we may not be successful.

4.4 PATHWAYS TO A RESILIENT FUTURE

In this final section we move beyond the general considerations of our method and its benefits and limitations to a short discussion of the main perspectives for future work arising from the three themes that comprised Chapter 3.

4.4.1 COMMON GOODS: TOWARD A SOVEREIGNTY OF RESOURCES?

In the first theme, "getting to know the territory," we grouped together several projects focused on the valorization of the cultural and natural heritage of the territory and the promotion of public spaces and common goods as the basis for the economic and cultural revitalization of rural areas.

The "green revolution," the name given to the transformation of agriculture using technological and science-based methods since the mid-20th century, is a key element of the global modernization project and has brought about enormous changes in traditional peasant societies. The green revolution has been widely criticized in recent years for its technocratic and industrializing tendencies. Shiva (2016, p. 21), for example, observed how the imposition of resource-intensive industrialized agriculture led to social breakdown in Punjab, India, while McMichael (2016) criticized the way in which development has come to mean replacing place-based farming with agroindustrialization. Other authors (Scoones & Thompson, 2011; Thompson, 2012) raised concerns over the privatization of genetically modified seeds by corporate seed breeders, such that farmers cannot save, breed, or exchange seeds and are obliged to purchase seed each year from the breeder. These problems have recently been most acutely felt in countries in the global South, but are by no means confined to them. While the technocratic model exemplified by the green revolution remains dominant, new social movements are emerging that seek to challenge it (Holt-Giménez & Altieri, 2013). Through this diverse range of organizations, which do not always have the same goals, the idea of a "sovereignty of resources" or common goods has begun to emerge. This idea simply states that each community should have the right to decide on the management of the lands, natural resources, and common goods in its locality. Though this has been the principle of effective governance of land and resources for many millennia, it is increasingly under threat by the global march toward privatization of commons (Goldman, 2001) and the tendency toward concentration of land in the hands of a small number of owners as a result of a liberalized land market.[8]

However, while the struggle for political change must continue until these aspects are addressed, there are many things than can be done at the local scale to prevent the disappearance or expropriation of important common goods like pastures, woodlands, roads, streams, meadows, etc. wherever they exist. But to ensure their survival, new uses must be found for them or traditional uses recovered. This is the case of the livestock droveways that formed the basis of several projects in Theme 3.1. The starting point of this work was the need to recover their use-values, either for the movement of livestock (transhumance and transterminance) or for other uses such as recreational trails.

On this basis, we propose three aspects as key to protecting and preserving **common goods** at the local scale.

- Value them, because if nobody cares about them, they will disappear.
- Propose new uses for them, recover traditional uses, and protect them from destruction and degradation.
- Defend the right to decide about their management at the local scale, establishing the local community as their principal stewards. Legal protection on its own is not sufficient to ensure the protection of commons if they are not adopted by the local community.

4.4.2 IMPORTANCE OF LINKS BETWEEN CITY AND COUNTRYSIDE

In Theme 3.2, "between city and country: building more resilient rural—urban relations," we look at how globalizing tendencies in general, and of food systems in

---

[8]"Where farmland is bought and sold like any other commodity and society allows the unlimited accumulation of farmland by a few, superfarms replace family farms and all of society suffers" (Rosset, Collins, & Lappe, 2000).

particular, have led urban areas to become disconnected from their surrounding regions. This has resulted in rural decline, degradation and abandonment of lands on the edge of the city, loss of cultural values associated with local food networks, and damage to ecosystem services and functions.

For years the city has lived with its back to the country, ignoring the problems of rural society (loss of cultural value of farming, low agricultural incomes, lack of generational replacement, poor public services, etc.), using rural areas alternately as waste dumps (out of sight, out of mind) or as leisure parks that exist solely for the pleasure of city dwellers and disappear once the weekend is over. Rural areas, in the collective imagination, tend to be idealized (beautiful, tranquil) or condemned (backward, uneconomic), with nothing in between. This is primarily because many rural areas, particularly in the rich world, have lost their function as food producers, having been outcompeted by countries where food can be produced more cheaply and transported long distances to consumers. By this logic a tomato coming from the other side of the world is just as good as one coming from the other side of the river, and even though the well-travelled tomato has emitted many more tonnes of carbon than the local one, its price may be lower.[9] As a result, many rural areas that traditionally supplied their nearest city have lost their productive economy entirely.

To restore this lost connection between city and country we need to work at multiple levels and scales. At the political level, we need to push for changes to our agribusiness models to favor local low-emission production over transcontinental networks of supply. But communities also need to work together to develop new strategies that restore and reinforce the mutual interdependence of urban and rural areas and promote a less confrontational and more respectful model of coexistence.[10] Society's perception of food is gradually changing, with urban populations becoming increasingly interested in good-quality local food with an emphasis on sustainable production and ethical treatment of livestock. This growing awareness among city dwellers is opening up new business models, like direct sale to the consumer at more favorable prices for the producer, removing intermediaries, and challenging the corporate power of the large food chains. At the same time, for rural areas to retain viable communities, provision of services cannot be left exclusively to the invisible hand of the market, since companies will invest where they stand to gain most clients. Overall, policy-makers need to be encouraged to think about the problem from the point of view of increasing society's resilience to global shocks rather than increasing gross domestic product.

---

[9]The consumer's greater understanding of food miles leads growers in cold countries to produce greenhouse crops for sale locally as more sustainable alternatives. Sometimes the carbon footprint of these local products, on account of the increased energy usage required to cultivate them, is greater than that of produce transported from overseas. Thus in the absence of a tax on carbon emissions, or any proper system of carbon accounting, some local products may be worse, not better, for the environment.
[10]By this we mean, for example, by putting the brakes on speculative urban development so farmlands do not become building lots in-waiting.

4.4.3 THE FUTURE OF COLLABORATIVE LAND AND RESOURCES PLANNING

In Theme 3.3, "conflicts, citizens and society: participatory modeling for a resilient future", we looked at how participatory processes can help tackle complex human–environment interaction problems, solve conflicts over land and resources, and find pathways to more sustainable futures.

The three cases presented showed a more detailed and interconnected range of participatory methods than the preceding two themes. The participatory process was transformed into a cycle of alternating analytical and discursive participatory modeling activities, during which the research team and the participants engaged repeatedly with each other, developed a rapport, and explored the limits of their original perspectives to open up new options. These very involved processes were only possible thanks to substantial research funding (from the Spanish government in one case and from the EU in the others). These types of project are slightly paradoxical, in the sense that they usually try to tackle problems that have defied solution through the ordinary land planning process and are thus by definition policy relevant (what could be more relevant to policy than finding out why policy is not working?), but at the same time they tend to be reflective, slow to develop, and sometimes also rather organic and hard to control (it is very difficult to know at the start whether you will be even moderately successful), which clearly situates them in the domain of research.

However, although land planners and regulating authorities have traditionally tended to work by applying procedures set down in law or guidance to specific cases that arise (generally through development applications), a task that usually gives no time to become involved in complicated knowledge-sharing activities like these, they are increasingly becoming involved in them nonetheless. Even where employers (usually a public body) do not openly acknowledge the need for their staff to play such a role, they often tolerate it.[11] This seems to indicate a positive shift in the way this sector thinks about land and resources planning in favor of a more consultative approach. However, the types of example we show are too time consuming, at present, to form part of the day-to-day decision-making process in planning as it is currently practiced. To be able to arrive at our main goal, a world in which land and resources planning decisions are habitually taken by communities themselves through careful, reflective participatory processes unrelated to electoral cycles, 5-year plans, or housing quotas, the already favorable inclination of planners themselves toward greater citizen participation could potentially be used to

---

[11]Yet we note that small, under-resourced bodies did not usually participate (we found it consistently difficult to engage planners from municipalities). Three principal reasons can be advanced: because budgets and staff time for non-priority activities were limited; because small municipalities tended to lack a staff member with the desired profile; and because politicians are much closer to the planners at the municipal level and can more easily block participation in initiatives they disapprove of than in a larger organization.

transform the system. Although this kind of transformation may be uncomfortable, its benefits, in terms of limiting conflict and increasing the probability of social acceptance and durability of interventions agreed, are likely to outweigh the disadvantages.

4.4.4 FROM DECISION SUPPORT TOOLS TO POLICY OPTION GENERATORS

The application of quantitative analytical techniques like geographical analysis or spatial modeling to socioenvironmental problems has its origins in the quantitative revolution of the second half of the 20th century. Though there is no doubt that quantitative approaches have revolutionized environmental science, they have also left a slightly technocratic legacy in the way we look at environmental decision-making. There remains an expectation, common to many policy-makers, planners, and academics, that solutions to complex problems necessarily require more tools, better data, and deeper analysis. In fact this is only one aspect of the many-sided wicked problem domain of environmental decision-making, and continued emphasis on only this aspect has led to, as one stakeholder we worked closely with argued, an "analytical—technical dictatorship" in environmental planning.

A key example of this relates to the role of complex environmental models like those we developed under TiGrESS, DUSPANAC and COMPLEX as so-called "decision support tools." This is a very frequently used term, which tends to give the impression that these models are reliable pieces of equipment that produce identical answers, like telescopes or surveying instruments that can be stored and transported to wherever they are needed. In fact, while the underlying algorithms on which the computational aspects depend are mathematically robust and very reliable, this tells us nothing about the utility or applicability of a model to a given situation. This is the reason why calibration is so important, and it is this which leads us to suggest that the best way to calibrate these models is through participatory processes. In the end, despite widespread expectation that these kinds of computational models would be directly used by policy-makers and their advisors to make decisions about the environment, this has not come to pass. In fact, these kinds of models are very rarely used outside of the domain of research. The expectation that they ever would or could be used in such a way gives us an insight both into the strong grip that the technocratic model of land governance holds over our imagination and the degree to which the real strengths and weaknesses of these models have been misunderstood. Thus rather than condemning society as "insufficiently scientific" to make proper use of these models, we need to take a step back and look more broadly at the contribution the models can really make to the goal of understanding and planning our environment.

Models have a wide range of potential uses, as Epstein[12] noted. Participatory environmental models like those presented in the case studies in Chapter 3, Theme 3.3, are intended to play a broader role than simply providing "a turn-key itinerary" (Barreteau et al., 2003). Among the roles these models have played in the work discussed in this book we can highlight just a few: to show how the many viewpoints of stakeholders lead to complex land-use outcomes; to show how the landscape evolves over time to produce unintended and unplanned consequences; to formalize and structure the different types of information that stakeholders bring to the table; to serve as a platform for discussion around frequently occurring issues of concern and disagreement; to visualize land change in a real location well known to the stakeholder group; to demonstrate and argue about the relevance of scientific analysis to land-planning decisions; to visualize the effect of current policy on future land use; to build and project a wide range of stakeholder views about what may happen in the future; to illustrate the failure of the "paradigm of endless growth" model of development; to explore policy actors' limits and boundaries; and to generate imaginative new options for managing the territory. There are doubtless many other purposes that these models can serve.

Thus the most important application of these models to policy is not as tools for stakeholders to pick up and use to make planning decisions, but as processes to develop and explore the problem domain alongside other key actors in the context of their different needs and interests. Decisions will of course have to be made, but not by pressing the "on" button on a decision support tool and waiting to see what comes out, because land planning never works in this way, but by reviewing and balancing options that have emerged collectively from the stakeholder community, and then negotiating to put them into practice. In this sense, these models are more "option generators" for a future territorial policy than "support tools for decision-making."

4.4.5 FROM PREDICTING THE FUTURE TO EXPLORING MULTIPLE PATHWAYS

Finally, for these kinds of model processes to fulfill their potential it is important that stakeholders understand their limitations. Models are always susceptible to be misinterpreted. Future simulations are often understood as predictions or prophecies of a single future option that has to be fulfilled. Modelers may confuse the issue by trying to explain uncertainty instead of simply emphasizing that future simulations are

---

[12]Epstein (2008) offers 16 reasons for building a model. 1. Explain. 2. Guide data collection. 3. Illuminate core dynamics. 4. Suggest dynamical analogies. 5. Discover new questions. 6. Promote a scientific habit of mind. 7. Bound (bracket) outcomes to plausible ranges. 8. Illuminate core uncertainties. 9. Offer crisis options in near-real time. 10. Demonstrate tradeoffs/suggest efficiencies. 11. Challenge the robustness of prevailing theory through perturbations. 12. Expose prevailing wisdom as incompatible with available data. 13. Train practitioners. 14. Discipline the policy dialogue. 15. Educate the general public. 16 Reveal the apparently simple (complex) to be complex (simple).

intended to explore the possible outcomes of observable tendencies, not predict a single future. The challenge is not so much in explaining that this is not the real goal of such a model, but in explaining why a model that does not do this is such a useful tool. Again, technocratic approaches to planning, which lead us to expect a single best answer to very complex problems, cast a long shadow. The approach we have advocated is to involve stakeholders gradually as participant modelers throughout the process, and to avoid suddenly producing computational models that only researchers understand as tools for solving these kinds of problems. Of course models sometimes are, necessarily, too complex for everyone to understand right away, but in these cases extra effort is needed to explain why they are necessary.

4.4.6 ENSURING LEGACY: THE IMPORTANCE OF NETWORKS

The empowerment of social actors as key agents of change, the fundamental purpose of this book, cannot be effective without parallel work on the development and maintenance of extensive social networks. Not only are these networks indispensable for undertaking participatory processes of the kind we have described, but they are also essential for putting the initiatives that emerge out of such processes into practice. We have explored a range of techniques for facilitating the legacy of actions agreed by stakeholder collectives, like objective trees, action plans, and working groups, but ultimately the success of an initiative as a catalyst for real change in the territory depends on the local participants, not on the facilitators, who must move on to new projects and challenges. But here we run into difficulty, because even the best-funded research project is not able to bear the cost of maintenance of the networks that the project has built for long after its completion. The success of participatory work carried out by organizations like our own, Observatorio para una Cultura del Territorio, depends entirely on networks that are built up over long periods of time. To move toward a new, more consultative model of land and resources planning, new models of financing will need to be found to address this issue.

4.4.7 THE IMPORTANCE OF SOCIAL TRANSFORMATION

Overcoming the serious environmental challenges that society is facing implies a social transformation. The relationship between policy-makers and citizens around land and resources governance in particular needs to change to facilitate this transformation. This changing relationship begets new interactions, giving rise to new channels of participation and engagement and new governance structures that respond to these changes. The global context, as we write, is very challenging, with undemocratic governments seemingly on the rise and democratic ones pushing discredited austerity economics on unwilling citizens and resisting calls for change. Under the very resilient neoliberal growth paradigm, big business is flexing its muscles at the expense of social justice and the environment, and many 20th-century problem-solving approaches and institutional arrangements are proving to be ineffective in the face of wicked problems like climate change. In this context, flexible democratic structures that are strong enough to limit the power of global elites and open enough to allow citizen-led transition processes to flourish are increasingly necessary. The future agenda for participatory environmental research should be directed toward improving information transfer and co-creation of knowledge among the major groups of social actors (business, policy, science, civil society). These days we are overwhelmed with data, but we lack adequate understanding of the interactions among social actors and institutions and we lack tools to facilitate peaceful social transformation to keep the impact of human societies within planetary boundaries. In spite of the apparent magnitude of this task, recognizing the need for social transformation is the first step. In the words of the poet Antonio Machado[13] "Traveller, there is no road to follow, the road is made by walking."

---

[13]Machado, 2016 [1907-17].

References

Alcamo, J. (2008). Chapter six the SAS approach: combining qualitative and quantitative knowledge in environmental scenarios. *Developments in Integrated Environmental Assessment, 2,* 123–150.

Alonso, P. M., Hewitt, R., Pacheco, J. D., Bermejo, L. R., Jiménez, V. H., Guillén, J. V., et al. (2016). Losing the roadmap: renewable energy paralysis in Spain and its implications for the EU low carbon economy. *Renewable Energy, 89,* 680–694.

Altieri, M. A. (2009). Agroecology, small farms, and food sovereignty. *Monthly Review, 61*(3), 102.

Alvial-Palavicino, C., Garrido-Echeverría, N., Jiménez-Estévez, G., Reyes, L., & Palma-Behnke, R. (2011). A methodology for community engagement in the introduction of renewable based smart microgrid. *Energy for Sustainable Development, 15*(3), 314–323.

Antunes, P., Santos, R., & Videira, N. (2006). Participatory decision making for sustainable development—the use of mediated modelling techniques. *Land Use Policy, 23*(1), 44–52.

Arnstein, S. R. (1969). A ladder of citizen participation. *Journal of the American Institute of planners, 35*(4), 216–224.

Atlee, T. (2014). *The tao of democracy: Using co-intelligence to create a world that works for all.* North Atlantic Books.

Bagstad, K. J., Reed, J. M., Semmens, D. J., Sherrouse, B. C., & Troy, A. (2016). Linking bio-physical models and public preferences for ecosystem service assessments: a case study for the Southern Rocky Mountains. *Regional Environmental Change, 16*(7), 2005–2018.

Balram, S., & Dragićević, S. (2005). Attitudes toward urban green spaces: integrating questionnaire survey and collaborative GIS techniques to improve attitude measurements. *Landscape and Urban Planning, 71*(2), 147–162.

Banerjee, S. B. (2000). Whose land is it anyway? National interest, indigenous stakeholders, and colonial discourses: the case of the Jabiluka uranium mine. *Organization & Environment, 13*(1), 3–38.

Barreteau, O., Antona, M., d'Aquino, P., Aubert, S., Boissau, S., Bousquet, F., et al. (2003). Our companion modelling approach. *Journal of Artificial Societies and Social Simulation, 6*(2).

Berg, B. L., Lune, H., & Lune, H. (2004). *Qualitative research methods for the social sciences* (Vol. 5). Boston, MA: Pearson.

Berkes, F., & Folke, C. (1998). Linking social and ecological systems for resilience and sustainability. *Linking Social and Ecological Systems: Management Practices and Social Mechanisms for Building Resilience, 1,* 13–20.

Berkes, F., Folke, C., & Colding, J. (2000). *Linking social and ecological systems: Management practices and social mechanisms for building resilience.* Cambridge University Press.

Blackstock, K. L., Kelly, G. J., & Horsey, B. L. (2007). Developing and applying a framework to evaluate participatory research for sustainability. *Ecological Economics, 60*(4), 726–742.

Borda, O. F. (2001). Participatory (action) research in social theory: Origins and challenges. In *Handbook of action research: Participative inquiry and practice* (pp. 27–37).

Boulos, M. N. K., Lu, Z., Guerrero, P., Jennett, C., & Steed, A. (2017). From urban planning and emergency training to Pokémon Go: applications of virtual reality GIS (VRGIS) and augmented reality GIS (ARGIS) in personal, public and environmental health. *International Journal of Health Geographics, 16*(1), 7.

Bourlakis, M., Maglaras, G., Aktas, E., Gallear, D., & Fotopoulos, C. (2014). Firm size and sustainable performance in food supply chains: insights from Greek SMEs. *International Journal of Production Economics, 152,* 112–130.

Brown, C., & Miller, S. (2008). The impacts of local markets: a review of research on farmers markets and community supported agriculture (CSA). *American Journal of Agricultural Economics, 90*(5), 1298–1302.

Brown, E., Dury, S., & Holdsworth, M. (2009). Motivations of consumers that use local, organic fruit and vegetable box schemes in Central England and Southern France. *Appetite, 53*(2), 183–188.

Brown, G., de Bie, K., & Weber, D. (2015). Identifying public land stakeholder perspectives for implementing place-based land management. *Landscape and Urban Planning, 139,* 1–15.

Brown, G., & Kyttä, M. (2014). Key issues and research priorities for public participation GIS (PPGIS): a synthesis based on empirical research. *Applied Geography, 46,* 122–136.

Brown, G., Weber, D., & de Bie, K. (2015). Is PPGIS good enough? An empirical evaluation of the quality of PPGIS crowd-sourced spatial data for conservation planning. *Land Use Policy, 43,* 228–238.

Brown, G. G., & Pullar, D. V. (2012). An evaluation of the use of points versus polygons in public participation geographic information systems using quasi-experimental design and Monte Carlo simulation. *International Journal of Geographical Information Science, 26*(2), 231–246.

Brown, K., Adger, W. N., Tompkins, E., Bacon, P., Shim, D., & Young, K. (2001). Trade-off analysis for marine protected area management. *Ecological Economics, 37*(3), 417–434.

Brundtland. (1987). *Our common future.* United Nations Organization.

Burgess, T. F. (2001). Guide to the design of questionnaires. In *A general introduction to the design of questionnaires for survey research* (pp. 1–27).

Carson, R. (2002 [1962]). *Silent spring.* Boton, USA: Houghton Mifflin.

Cartledge, K., Dürwächter, C., Hernandez-Jimenez, V., & Winder, N. (2009). Making sure you solve the right problem. *Ecology and Society, 14*(2), R3. http://www.ecologyandsociety.org/vol14/iss2/resp3/.

Carver, S. J. (1991). Integrating multi-criteria evaluation with geographical information systems. *International Journal of Geographical Information System, 5*(3), 321–339.

Cavallaro, F. (2009). Multi-criteria decision aid to assess concentrated solar thermal technologies. *Renewable Energy, 34*(7), 1678–1685.

CBI Market Intelligence. (2015). *CBI trade statistics: Fresh fruit and vegetables in Europe.* Available at https://www.cbi.eu/sites/default/files/market_information/researches/trade-statistics-europe-fresh-fruit-vegetables-2015.pdf.

Cembranos, F., & Medina, J. A. (2003). *Grupos Inteligentes: Teoría y práctica del trabajo en equipo.* Editorial Popular.

Chambers, R. (1994a). Participatory rural appraisal (PRA): analysis of experience. *World Development, 22*(9), 1253–1268.

Chambers, R. (1994b). The origins and practice of participatory rural appraisal. *World Development, 22*(7), 953–969.

Chambers, R. (1997). *Whose reality counts? Putting the first last.* Londres: Intermediate Technology Publications.

Cheng, S., Chan, C. W., & Huang, G. H. (2003). An integrated multi-criteria decision analysis and inexact mixed integer linear programming approach for solid waste management. *Engineering Applications of Artificial Intelligence, 16*(5), 543–554.

Conde, F. (1993). *Los métodos extensivos e intensivos en la investigación social de las drogodependencias. VVAA Las drogodependencias: Perspectivas sociológicas actuales* (pp. 203–230). Madrid: Colegio de Sociólogos.

Coulthard, M., & Coulthard, M. (2014). *An introduction to discourse analysis*. Routledge.

Creswell, J. W., & Clark, V. L. P. (2007). *Designing and conducting mixed methods research*.

Cueto, L.Á. C., & Muñoz, S. F. (2016). Movimientos sociales y políticas urbanas locales en tiempo de crisis: el caso de Torrelodones. *Ciudad y territorio: Estudios territoriales, 188*, 261–279.

Daily, G. C. (Ed.). (1997). *Nature's services: societal dependence on natural ecosystems*. Washington, DC: Island Press.

Daley, P. J., & James, B. L. (1992). Ethnic broadcasting in Alaska: the failure of a participatory model. *Ethnic Minority Media: An International Perspective, 13*, 23.

De Boer, C., & Bressers, H. (June 2011). Contextual interaction theory as a conceptual lens on complex and dynamic implementation process. In *Proceedings of conference: challenges of making public administration and complexity theory work*.

De Castella, K., McGarty, C., & Musgrove, L. (2009). Fear appeals in political rhetoric about terrorism: an analysis of speeches by Australian Prime Minister Howard. *Political Psychology, 30*(1), 1–26.

De Kraker, J., & van der Wal, M. (2012). How to make environmental models better in supporting social learning. In *Proceedings of the 2012 international congress on environmental modelling and software: managing resources of a limited planet, Leipzig, Germany*.

Devine-Wright, P. (2007). *Reconsidering public attitudes and public acceptance of renewable energy technologies: A critical review*. Manchester: School of Environment and Development, University of Manchester. Available at http://www.sed.manchester.ac.uk/research/beyond_nimbyism.

el Diario. (2017). *El fuego entra en el Espacio Natural de Doñana* [*The fire enters the Doñana natural area*], 25th June 2017 http://www.diariodehuelva.es/2017/06/25/fuego-entra-espacio-natural-donana/.

Doulton, H., & Brown, K. (2009). Ten years to prevent catastrophe?: Discourses of climate change and international development in the UK press. *Global Environmental Change, 19*(2), 191–202.

Ehrlich, P. (1968). *The population bomb*. USA: Sierra Club/Ballantine Books.

EME. (2011). *Evaluación de Ecosistemas del Milenio en España. Conservación de los Servicios de los Ecosistemas y la biodiversidad para el bienestar humano* [*Millennium ecosystems evaluation in Spain. Conservation of ecosystem services and biodiversity for human wellbeing.*]. Madrid: Madrid Autonomous University. Final report.

Encinas Escribano, A. M., & Winder (coordinators), N. (2005). Hacia una planificación territorial sostenible en la Comunidad de Madrid: Directrices y Recomendaciones. *Policy Briefing for the TiGrESS Project*.

Engelen, G., Lavalle, C., Barredo, J. I., Meulen, M. V. D., & White, R. (2007). The MOLAND modelling framework for urban and regional land-use dynamics. *Modelling Land-Use Change*, 297–319.

Epstein, J. M. (2008). Why model? *Journal of Artificial Societies and Social Simulation, 11*(4), 12.

Escobar, F., Hewitt, R. J., & Hernández Jiménez, V. (2016). *Usos del suelo en los parques nacionales españoles. Evolución y modelado participativo, Proyectos de investigación en parques nacionales* (pp. 2010–2013). Madrid: OAPN.

Étienne, M. (2011). *Companion modelling. A participatory approach to support sustainable development*. Versailles: Quae.

European Commission (EC). (2015). *Report from the commission to the European parliament, the council, the European economic and social committee and the committee of the regions*. Renewable energy progress report, SWD (2015) 117 final. Unpublished EC Report. Available from https://ec.europa.eu/transparency/regdoc/rep/1/2015/EN/1-2015-293-EN-F1-1.PDF.

Fagerholm, N., Oteros-Rozas, E., Raymond, C. M., Torralba, M., Moreno, G., & Plieninger, T. (2016). Assessing linkages between ecosystem services, land-use and well-being in an agroforestry landscape using public participation GIS. *Applied Geography, 74*, 30−46.

Fairclough, N. (2013). *Critical discourse analysis: The critical study of language*. Routledge.

Farinós, J., Romero, J., & Salom (coords.), J. (2009). *Cohesión e inteligencia territorial. Dinámicas y procesos para una mejor planificación y toma de decisiones*. Valencia: Publicaciones de la Universidad de Valencia.

Farrington, J., & Martin, A. M. (1988). Farmer participatory research: a review of concepts and recent fieldwork. *Agricultural Administration and Extension, 29*(4), 247/264.

Feindt, P. H., & Oels, A. (2005). Does discourse matter? Discourse analysis in environmental policy making. *Journal of Environmental Policy & Planning, 7*(3), 161−173.

Flacke, J., & de Boer, C. (2016). An interactive GIS-tool for collaborative local renewable energy planning. In *AGILE conference, Wageningen 2016*. https://agile-online.org/conference_paper/cds/agile_2016/shortpapers/130_Paper_in_PDF.pdf.

FM (2016). Website of the Moderna Foundation. Accessed August 2016. http://www.modernanavarra.com/.

Friedland, W. H. (1994). The global fresh fruit and vegetable system: an industrial organization analysis. *The Global Restructuring of Agro-Food Systems*, 173−189.

Gaddis, E. J. B., Vladich, H., & Voinov, A. (2007). Participatory modeling and the dilemma of diffuse nitrogen management in a residential watershed. *Environmental Modelling & Software, 22*(5), 619−629.

Gasper, D. (2000). Evaluating the 'logical framework approach' towards learning-oriented development evaluation. *Public Administration & Development, 20*(1), 17.

Gee, J. P. (2014). *An introduction to discourse analysis: Theory and method*. Routledge.

Geilfus, F. (2002). *Guía Metodológica para el manejo de conflictos ambientales y de recursos naturales*. Costa Rica: IICA.

Geilfus, F. (2008). *80 tools for participatory development: Appraisal, planning, follow-up and evaluation* (No. 303.4 G312e). San José, Costa Rica.

Gobierno de Canarias. (2015). *Informe Estadístico del Mapa de Cultivos de Canarias* [*Statistical report on the map of crops in the Canary Islands*] (p. 29). Servicio de Planificación de Obras y Ordenación Rural. Consejería de Agricultura, Ganadería, Pesca y Aguas. Santa Cruz de Tenerife.

Goldman, M. (2001). *Privatising nature: Political struggles for the global commons, 1998*. London: Pluto.

Gomez Orea, D. (2001). *Ordenacion territorial. Ediciones MundiPrensa/Editorial Agricola Española*. Madrid: España.

Gudynas, E. (2011). Buen vivir: Todays' tomorrow. *Development, 54*(4), 441−447. http://dx.doi.org/10.1057/dev.2011.86.

Guzmán Casado, G., Gonzalez de Molina, M., & Sevilla Guzman, E. (2000). *Introducción a la agroecología como desarrollo rural sostenible*. Madrid: MundiPrensa.

Hajer, M., & Versteeg, W. (2005). A decade of discourse analysis of environmental politics: achievements, challenges, perspectives. *Journal of Environmental Policy & Planning, 7*(3), 175–184.

Hammersley, M. (2014). On the ethics of interviewing for discourse analysis. *Qualitative Research, 14*(5), 529–541.

Hardin, G. (1968). The tragedy of the commons. *Science, 162*, 1243–1248.

Hayek, F. A. (2014). The road to serfdom. In *Text and documents: The definitive edition* (Vol. 2). Routledge.

Healey, P. (1994). *Collaborative planning: Shaping places and fragmented societies.* London: McMillan.

Heras López, M. (2015). *Towards new forms of learning. Exploring the potential of participatory theatre in sustainability science* (Ph.D. thesis). University of Barcelona.

Hernández Jiménez, V. (2006). *Stakeholder engagement: The Madrid experience.* Unpublished report for the TiGrESS project, WP5.

Hernández-Jiménez, V. (2007). Participatory land planning in the region of Madrid, Spain: An integrative perspective. (Unpublished Ph.D. thesis), Newcastle University, Newcastle upon Tyne, UK.

Hernandez-Jimenez, V., Ocon, B., Encinas, M. A., Pereira, D., & Winder, N. (2009). Planificacion territorial participativa en el entorno de las grandes ciudades: Madrid y sus relaciones urbano-rurales. In J. Farinos, J. Romero, & J. Salomon (Eds.), *Cohesion e Inteligencia Territorial: Dinamicas y procesos para una mejor planificacion y toma de decisiones.* Universitat de Valencia.

Hernández-Jimenez, V., Ocón, B., & Vicente, J. (2009). *Espacios periurbanos: Transición de la ciudad al campo. La importancia de los espacios agrarios en las grandes ciudades.* Revista ECOSOSTENIBLE. Valencia: CISS.

Hernández Jiménez, V., & Winder, N. (2006). *Running experiments with the Madrid simulation model.* TiGrESS Final report http://citeseerx.ist.psu.edu/viewdoc/download?doi=10.1.1.119.9495&rep=rep1&type=pdf.

Hernandez-Jimenez, V., & Winder, N. (2009). Chapter 12 knowledge integration and power relations: pathways to sustainability in Madrid. In *Beyond the rural-urban divide: cross-continental perspectives on the differentiated countryside and its regulation* (pp. 305–322). Emerald Group Publishing Limited.

Hewitt, R., de Boer, C., Pacheco, J. D., Hernández Jiménez, V., Alonso, P. M., Román, L., et al. (2015). *APoLUS model full system documentation.* EU FP7 COMPLEX project report, deliverable D3.5. Available from http://owsgip.itc.utwente.nl/projects/complex/images/uploaded_files/WP3_deliverable_3.5_09112015_errors_corrected.pdf.

Hewitt, R., & Díaz-Pacheco, J. (2017). Stable models for metastable systems? Lessons from sensitivity analysis of a cellular automata urban land use model. *Computers, Environment and Urban Systems, 62*, 113–124.

Hewitt, R., & Escobar, F. (2011). The territorial dynamics of fast-growing regions: Unsustainable land use change and future policy challenges in Madrid, Spain. *Applied Geography, 31*(2), 650–667.

Hewitt, R., & Hernández Jiménez, V. (2016). Towards a low-carbon future in Spain: policy briefing. Unpublished report of the COMPLEX project: D3.7. Available at https://www.researchgate.net/publication/308515473_Towards_a_low-carbon_future_in_Spain_Policy_briefing.

Hewitt, R., & Hernandez-Jimenez, V. (2010). Devolved regions, fragmented landscapes: the struggle for sustainability in Madrid. *Sustainability, 2*(5), 1252–1281.

Hewitt, R., Hernández Jiménez, V., Román Bermejo, L., & Escobar, F. (2017). Who knows best? The role of stakeholder knowledge in land use models- an example from Doñana, SW Spain. In M. T. Camacho Olmedo, M. Paegelow, J. F. Mas, & F. Escobar (Eds.), *Lecture notes in geoinformation and cartography' LNGC seriesGeomatic simulations and scenarios for modelling LUCC: A review and comparison of modelling techniques*. Springer. http://www.springer.com/series/7418.

Hewitt, R., Jiménez, V. H., Navarro, M. L., de la Cruz Lecanda, A., & Escobar, F. (2016). ¿Qué futuro queremos? Generación de indicadores medioambientales para escenarios de futuro. In *XVII Congreso Nacional de Tecnologías de Información Geográfica, Málaga, 29-30th June and 1 of July 2016* (pp. 155–164). Available at http://www.age-geografia.es/tig/2016_Malaga/Hewitt.pdf.

Hewitt, R., Pera, F., & Escobar, F. (2016). *Cambios recientes en la ocupación del suelo de los parques nacionales españoles y su entorno*. Cuadernos Geográficos de la Universidad de Granada.

Hewitt, R., Van Delden, H., & Escobar, F. (2014). Participatory land use modelling, pathways to an integrated approach. *Environmental Modelling & Software, 52*, 149–165.

Hewitt, R. J., Winder, N. P., Jiménez, V. H., Alonso, P. M., & Bermejo, L. R. (2017). Innovation, pathways and barriers in Spain and beyond: An integrative research approach to the clean energy transition in Europe. *Energy Research & Social Science, 34*, 260–271.

Higgs, G. (2006). Integrating multi-criteria techniques with geographical information systems in waste facility location to enhance public participation. *Waste Management & Research, 24*(2), 105–117.

Hill, T., & Westbrook, R. (1997). SWOT analysis: it's time for a product recall. *Long Range Planning, 30*(1), 46–52.

Hobbs, B. F., & Horn, G. T. (1997). Building public confidence in energy planning: a multi-method MCDM approach to demand-side planning at BC gas. *Energy Policy, 25*(3), 357–375.

Holling, C. S., & Meffe, G. K. (1996). Command and control and the pathology of natural resource management. *Conservation Biology, 10*(2), 328–337.

Holt-Giménez, E., & Altieri, M. A. (2013). Agroecology, food sovereignty, and the new green revolution. *Agroecology and Sustainable Food Systems, 37*(1), 90–102.

Homer-Dixon, T., Walker, B., Biggs, R., Crépin, A. S., Folke, C., Lambin, E., et al. (2015). Synchronous failure: the emerging causal architecture of global crisis. *Ecology and Society, 20*(3).

INE (2003). Madrid: Land Registry.

Ite, U. E. (1996). Community perceptions of the Cross River national park, Nigeria. *Environmental Conservation, 23*(04), 351–357.

Jones, N. A., Perez, P., Measham, T. G., Kelly, G. J., d'Aquino, P., Daniell, K. A., et al. (2009). Evaluating participatory modeling: developing a framework for cross-case analysis. *Environmental Management, 44*(6), 1180.

Kemp, R., & Martens, P. (2007). Sustainable development: how to manage something that is subjective and never can be achieved? *Sustainability: Science, Practice, & Policy, 3*(2), 5–14.

Kemp, R., Schot, J., & Hoogma, R. (1998). Regime shifts to sustainability through processes of niche formation: the approach of strategic niche management. *Technology Analysis & Strategic Management, 10*(2), 175–198.

Kesby, M. (2000). Participatory diagramming: deploying qualitative methods through an action research epistemology. *Area, 32*(4), 423–435.

Kolinjivadi, V., Gamboa, G., Adamowski, J., & Kosoy, N. (2015). Capabilities as justice: Analysing the acceptability of payments for ecosystem services (PES) through 'social multi-criteria evaluation'. *Ecological Economics, 118*, 99–113.

Kontogianni, A., Skourtos, M. S., Langford, I. H., Bateman, I. J., & Georgiou, S. (2001). Integrating stakeholder analysis in non-market valuation of environmental assets. *Ecological Economics, 37*(1), 123–138.

Kowalski, K., Stagl, S., Madlener, R., & Omann, I. (2009). Sustainable energy futures: Methodological challenges in combining scenarios and participatory multi-criteria analysis. *European Journal of Operational Research, 197*(3), 1063–1074.

Krueger, T., Page, T., Hubacek, K., Smith, L., & Hiscock, K. (2012). The role of expert opinion in environmental modelling. *Environmental Modelling & Software, 36*, 4–18.

Kuzemko, C. (2016). Energy depoliticisation in the UK: Destroying political capacity. *The British Journal of Politics and International Relations, 18*(1), 107–124.

Lee, A. H., Chen, H. H., & Kang, H. Y. (2009). Multi-criteria decision making on strategic selection of wind farms. *Renewable Energy, 34*(1), 120–126.

Lemon, M., Seaton, R., & Park, J. (1994). Social enquiry and the measurement of natural phenomena: the degradation of irrigation water in the Argolid Plain, Greece. *The International Journal of Sustainable Development & World Ecology, 1*(3), 206–220.

Leopold, A. (1949). *A sand county almanac*. Oxford University Press.

Lomborg, B. (2001). The truth about the environment. *The Economist, 360*(8233), 63–65.

Longhurst, R. (2003). Semi-structured interviews and focus groups. *Key Methods in Geography*, 117–132.

Lorenz, E. (1972). *Predictability: Does the flap of a butterfly's wing in Brazil set off a tornado in Texas?*. Available from http://gymportalen.dk/sites/lru.dk/files/lru/132_kap6_lorenz_artikel_the_butterfly_effect.pdf.

Lupton, D. (1992). Discourse analysis: a new methodology for understanding the ideologies of health and illness. *Australian Journal of Public Health, 16*(2), 145–150.

Machado, A. (2016). Proverbios y cantares (Campos de Castilla), XXIX [1907-17] Accessed: October 2017 https://es.wikisource.org/wiki/Proverbios_y_cantares_(Campos_de_Castilla).

Macleod, C. K., Blackstock, K., & Haygarth, P. (2008). Mechanisms to improve integrative research at the science-policy interface for sustainable catchment management. *Ecology and Society, 13*(2).

Mapedza, E., Wright, J., & Fawcett, R. (2003). An investigation of land cover change in Mafungautsi Forest, Zimbabwe, using GIS and participatory mapping. *Applied Geography, 23*(1), 1–21.

Marchioni, M. (1994). In *La utopía posible: La intervención comunitaria en las nuevas condiciones sociales*. Tenerife: Becnhomo.

Martín-López, B., García-Llorente, M., Palomo, I., & Montes, C. (2011). The conservation against development paradigm in protected areas: valuation of ecosystem services in the Doñana social–ecological system (southwestern Spain). *Ecological Economics, 70*(8), 1481–1491.

Martínez-Alier, J., Pascual, U., Vivien, F. D., & Zaccai, E. (2010). Sustainable de-growth: mapping the context, criticisms and future prospects of an emergent paradigm. *Ecological Economics, 69*(9), 1741–1747.

McGrath, J. E. (1984). *Groups: Interaction and performance* (Vol. 14). Englewood Cliffs, NJ: Prentice-Hall.

McIntyre, A. (2007). *Participatory action research* (Vol. 52). Sage Publications.

McMichael, P. (2016). Food security, land, and development. In *The palgrave handbook of international development* (pp. 671–693). UK: Palgrave Macmillan.

Meadows, D. H., Meadows, D. L., Randers, J., & Behrens, W. (1972). *The limits to growth. A report for the club of rome's project on the predicament of mankind.* New York: Potomac Associates.

Medley, K. E., Okey, B. W., Barrett, G. W., Lucas, M. F., & Renwick, W. H. (1995). Landscape change with agricultural intensification in a rural watershed, southwestern Ohio, USA. *Landscape Ecology, 10*(3), 161–176.

Midgley, G., Munlo, I., & Brown, M. (1998). The theory and practice of boundary critique: developing housing services for older people. *Journal of the Operational Research Society, 49*, 467–478.

Moffat, K., & Zhang, A. (2014). The paths to social licence to operate: an integrative model explaining community acceptance of mining. *Resources Policy, 39*, 61–70.

Montasell, J. (2006a). El repte de l'espai agrari periurbà. In *LA RELLA, revista digital del Congrés del Món Rural (Rural '06).*

Montasell, J. (2006b). Els espais agraris de la regió metropolitana de Barcelona. In *L'Atzavara núm. 14.*

Morán Alonso, N. (2015). *Dimensión territorial de los sistemas alimentarios locales. El caso de Madrid* [*the territorial dimension of local food systems, the case of Madrid. (Spanish, conclusions in English)*] (Ph.D. thesis). Madrid Polytechnic University.

Moratalla, A. Z. (2015). *El Parque Agrario: Estructura de preservación de los espacios agrarios en entornos urbanos en un contexto de cambio global* [*The agrarian park, structure of preservation for agrarian spaces en urban areas in the context of global change. (Spanish, conclusions in Italian)*] (Ph.D. thesis). Madrid Polytechnic University.

Narayanasamy, N. (2009). *Participatory rural appraisal: Principles, methods and application.* SAGE Publications India.

Naredo, J. M. (2008). El aterrizaje inmobiliario [the real estate crash-landing]. *Boletín CF+ S, 35.*

Naredo, J. M. (2009). Economía y poder. Megaproyectos, recalificaciones y contratas. In F. Aguilera, & J. M. Naredo (Eds.), *Economía, poder y megaproyectos, Lanzarote, Fundación César Manrique, Col. "Economía & Naturaleza".*

New York Times. (2011). *Grass roots party sways Spanish politics.* By Raphael Minder. July 26th 2011, Available at http://www.nytimes.com/2011/07/27/world/europe/27iht-madrid27.html.

Nielsen, K., Fredslund, H., Christensen, K. B., & Albertsen, K. (2006). Success or failure? Interpreting and understanding the impact of interventions in four similar worksites. *Work & Stress, 20*(3), 272–287.

Nikolaou, I. E., & Evangelinos, K. I. (2010). A SWOT analysis of environmental management practices in Greek mining and mineral industry. *Resources Policy, 35*(3), 226–234.

Northumberland National Park Historic Village Atlas: Alwinton, *Northumberland: An archaeological and historical study of a border township.* Available at: http://www.dartmoor.gov.uk/__data/assets/pdf_file/0005/147785/alwintonhva.pdf.

O'Riordan. (1998). Indicators for sustainable development. In *Proceedings of the European Commission (environment and climate programme) advanced study course 5th–12th July 1997, Delft, The Netherlands.*

Obermeyer, N. J. (1998). The evolution of public participation GIS. *Cartography and Geographic Information Systems, 25*(2), 65–66.

Ochoa, C. Y., & Moratalla, A. Z. (2015). *El Parque Agrario: Una figura de transición hacia nuevos modelos de gobernanza territorial y alimentaria.* Madrid: Heliconia S. Coop. Mad.

ONU. (1987). *Our common future. Informe Brundtland. Informe de la Comision Mundial sobre el Medioambiente y el Desarrollo.* Naciones Unidas.

Ostrom, E. (1990). *Governing the commons: The evolution of institutions for collective action.* Cambridge University Press.

Oteros-Rozas, E., Martín-López, B., Daw, T., Bohensky, E. L., Butler, J., Hill Martin-Ortega, J., et al. (2015). Participatory scenario planning in place-based social-ecological research: insights and experiences from 23 case studies. *Ecology and Society, 20*(4), 32. http://dx.doi.org/10.5751/ES-07985−200432.

Ottmann, G. (2005). *Agroecología y sociología histórica desde Latinoamérica.* Córdoba: Servicio de Publicaciones de la Universidad de Córdoba.

Owen, J. R., & Kemp, D. (2013). Social licence and mining: a critical perspective. *Resources Policy, 38*(1), 29−35.

Oxley, T., Jeffrey, P., & Lemon, M. (2002). Policy relevant modelling: relationships between water, land use, and farmer decision processes. *Integrated Assessment, 3*(1), 30−49.

Oxley, T., & Lemon, M. (2003). From social-enquiry to decision support tools: towards an integrative method in the Mediterranean rural environment. *Journal of Arid Environments, 54*(3), 595−617.

O'Brien, N., & Moules, T. (2007). So round the spiral again: a reflective participatory research project with children and young people. *Educational Action Research, 15*(3), 385−402.

de Pablo Valenciano, J., & Mesa, J. P. (2004). The competitiveness of Spanish tomato export in the European Union. *Spanish Journal of Agricultural Research, 2*(2), 167−180.

Pacheco, J. D., & Hewitt, R. (2010). *El territorio como bien de consumo: Las grandes superficies comerciales en el contexto metropolitano y su implicación para el desarrollo urbano sostenible [The territory as a consumer product: large shopping centres in their metropolitan context and their implication for sustainable development.]* (pp. 234−249). In Ciudad, territorio y paisaje: Reflexiones para un debate multidisciplinar.

Palomo, I. (2012). *El sistema socio-ecológico de Doñana ante el cambio global: Planificación de escenarios de eco-futuro.* Laboratorio de Socio-Ecosistemas de la Universidad Autónoma de Madrid.

Palomo, I., Martín-López, B., López-Santiago, C., & Montes, C. (2011). Participatory scenario planning for protected areas management under the ecosystem services framework: the Doñana social- ecological system in southwestern Spain. *Ecology and Society, 16*(1), 23.

Palomo, I., Martín-López, B., Zorrilla-Miras, P., del Amo, D. G., & Montes, C. (2014). Deliberative mapping of ecosystem services within and around Doñana National Park (SW Spain) in relation to land use change. *Regional Environmental Change, 14*(1), 237−251.

Pérez Orozco, A. (2014). *Subsersion feminista de la econmia, Aportes para un debate sobre el conflicto capital-vida.* Madrid: Traficantes de Sueños.

Piercy, N., & Giles, W. (1989). Making SWOT analysis work. *Marketing Intelligence & Planning, 7*(5/6), 5−7.

Plata-Rocha, W., Gómez-Delgado, M., & Bosque-Sendra, J. (2011). Simulating urban growth scenarios using GIS and multicriteria analysis techniques: a case study of the Madrid region, Spain. *Environment and Planning B: Planning and Design, 38*(6), 1012−1031.

Popper. (2005 [1935]). *The logic of scientific discovery.* Routledge.

Prell, C., Hubacek, K., Reed, M., Quinn, C., Jin, N., Holden, J., et al. (2007). If you have a hammer everything looks like a nail: traditional versus participatory model building. *Interdisciplinary Science Reviews, 32*(3), 263–282.

Raynolds, L. T., Murray, D., & Wilkinson, J. (Eds.). (2007). *Fair trade: The challenges of transforming globalization.* Routledge.

Reed, M., Evely, A. C., Cundill, G., Fazey, I. R. A., Glass, J., Laing, A., et al. (2010). What is social learning? *Ecology and Society, 15*(4), r1. http://www.ecologyandsociety.org/vol15/iss4/resp1/.

Reed, M. S., Graves, A., Dandy, N., Posthumus, H., Hubacek, K., Morris, J., et al. (2009). Who's in and why? A typology of stakeholder analysis methods for natural resource management. *Journal of Environmental Management, 90*(5), 1933–1949.

Reuters. (2016). *Vattenfall sells German lignite assets to Czech EPH, 18th April 2016.* http://www.reuters.com/article/us-vattenfall-germany-lignite-idUSKCN0XF1DV.

RIKS (2012). Metronamica Documentation, unpublished software manual distributed by the Research Institute for Knowledge Systems (RIKS).

Rogers, J. C., Simmons, E. A., Convery, I., & Weatherall, A. (2008). Public perceptions of opportunities for community-based renewable energy projects. *Energy Policy, 36*(11), 4217–4226.

Román Bermejo, L. P. (2016). *Metodologías Participativas para el Desarrollo Rural, Un enfoque desde la Agroecología* (Tesis doctoral). Universidad Internacional de Andalucía.

Romero, J., & Farinós, J. (2011). Redescubriendo la gobernanza más allá del buen gobierno. Democracia como base, desarrollo territorial como resultado. *Boletín de la Asociación de Geógrafos Españoles, 56*, 295–319.

Romero, J., Jiménez, F., & Villoria, M. (2012). (Un) Sustainable territories: causes of the speculative bubble in Spain (1996–2010) and its territorial, environmental, and sociopolitical consequences. *Environment and Planning C: Government and Policy, 30*(3), 467–486.

Rosset, P., Collins, J., & Lappe, F. M. (2000). Lessons from the green revolution. *Third World Resurgence*, 11–14.

Ruíz Romero, S., Santos, A. C., & Gil, M. A. C. (2012). EU plans for renewable energy. An application to the Spanish case. *Renewable Energy, 43*, 322–330.

Ruiz Ruiz, J. (2012). El grupo triangular: reflexiones metodológicas en torno a dos experiencias de investigación. *EMPIRIA. Revista de Metodología de las Ciencias Sociales, 24*.

Salgado, P. P., Quintana, S. C., Pereira, A. G., del Moral Ituarte, L., & Mateos, B. P. (2009). Participative multi-criteria analysis for the evaluation of water governance alternatives. A case in the Costa del Sol (Malaga). *Ecological Economics, 68*(4), 990–1005.

Sayago, S. (2015). The construction of qualitative and quantitative data using discourse analysis as a research technique. *Quality & Quantity, 49*(2), 727–737.

Scarpa, R., & Willis, K. (2010). Willingness-to-pay for renewable energy: Primary and discretionary choice of British households' for micro-generation technologies. *Energy Economics, 32*(1), 129–136.

Schultze, U., & Avital, M. (2011). Designing interviews to generate rich data for information systems research. *Information and Organization, 21*(1), 1–16.

Schumacher, E. F. (2011 [1973]). *Small is beautiful: A study of economics as if people mattered.* Random House.

Schwartz, S. H. (2003). A proposal for measuring value orientations across nations. *Questionnaire Package of the European Social Survey*, 259–290.

Schwartz, S. H. (2012). An overview of the Schwartz theory of basic values. *Online Readings in Psychology and Culture, 2*(1), 11.

Scoones, I., & Thompson, J. (2011). The politics of seed in Africa's green revolution: Alternative narratives and competing pathways. *ids Bulletin, 42*(4), 1–23.

Scottish Executive. (2006). *Sustainable development: A review of international literature. The centre for sustainable development* (pp. 1–177). University of Westminster and the Law School, University of Strathclyde.

Sepúlveda Ruiz, M., Calderón-Almendros, I., & Torres-Moya, F. J. (2012). *De lo individual a lo estructural. La investigación-acción participativa como estrategia educativa para la transformación personal y social en un centro de intervención con menores infractores* [*From the Individual to the Structural. Participatory Action Research as an Educational Strategy for Personal and Social Change in a Juvenile Offenders Centre*]. Ministerio de Educación.

Sheppard, S. R., Shaw, A., Flanders, D., Burch, S., Wiek, A., Carmichael, J., et al. (2011). Future visioning of local climate change: a framework for community engagement and planning with scenarios and visualisation. *Futures, 43*(4), 400–412.

Shiva, V. (2016). *The violence of the green revolution: Third world agriculture, ecology, and politics.* University Press of Kentucky.

Sieber, R. (2006). Public participation geographic information systems: A literature review and framework. *Annals of the Association of American Geographers, 96*(3), 491–507.

Simon, M., Zazo, A., Moran, N., & Hernandez-Jimenez, V. (2014). Pathways towards the integration of periurban agrarian ecosystems into the spatial planning system. *Ecological Processes, 3*(1), 1–16.

Skippers and Nicholson. (2011). *What are the implications of a 'presumption in favour of sustainable development'?*. http://www.annskippers.co.uk/wp-content/uploads/2011/04/What-are-the-implications-of-a-presumption-in-favour-of-sustainable-development.pdf.

Srivastava, P. K., Kulshreshtha, K., Mohanty, C. S., Pushpangadan, P., & Singh, A. (2005). Stakeholder-based SWOT analysis for successful municipal solid waste management in Lucknow, India. *Waste Management, 25*(5), 531–537.

Telemadrid. (2003). *Comments of Esperanza Aguirre made on 17th October 2003 in an interview, in reference to her party's proposals to create a map of the Community of Madrid defining urbanizable and protected areas.*

Terrados, J., Almonacid, G., & Hontoria, L. (2007). Regional energy planning through SWOT analysis and strategic planning tools: impact on renewables development. *Renewable and Sustainable Energy Reviews, 11*(6), 1275–1287.

Thayer, R. L. (2003). *LifePlace: Bioregional thought and practice.* Univ of California Press.

Thompson, C. B. (2012). Alliance for a green revolution in Africa (AGRA): advancing the theft of African genetic wealth. *Review of African Political Economy, 39*(132), 345–350.

Tobler, W. (1970). A computer movie simulating urban growth in the Detroit region. *Economic Geography, 46*(2), 234–240.

Toledo, V. M., & Barrera-Bassols, N. (2009). *La memoria biocultural. La importancia ecológica de las sabidurías tradicionales.* Barcelona: ICARIA.

Trakolis, D. (2001). Local people's perceptions of planning and management issues in Prespes Lakes National Park, Greece. *Journal of Environmental Management, 61*(3), 227–241.

Tress, B., & Tress, G. (2003). Scenario visualisation for participatory landscape planning—a study from Denmark. *Landscape and Urban Planning, 64*(3), 161–178.

Tress, B., Tress, G., & Fry, G. (2005a). *Defining concepts and the process of knowledge production in integrative research* (pp. 13–26). Heidelberg, Germany: Springer.

Tress, B., Tress, G., & Fry, G. (2005b). Integrative studies on rural landscapes: policy expectations and research practice. *Landscape and Urban Planning, 70*(1), 177–191.

Tress, G., Tress, B., & Van der Valk, A. (2003). Interdisciplinary and transdisciplinary landscape studies: Potential and limitations. Delta program. In *Delta series 2*. Wageningen: Wageningen University.

Tyrväinen, L., Mäkinen, K., & Schipperijn, J. (2007). Tools for mapping social values of urban woodlands and other green areas. *Landscape and Urban Planning, 79*(1), 5–19.

UN. (2005). *United Nations population division — world population prospects: The revision and world urbanization prospects: The 2005 revision*. New York: United Nations.

Van Delden, H., Engelen, G., Uljee, I., Hagen, A., Van der Meulen, M., & Vanhout, R. (2005). *PRELUDE quantification and spatial modelling of land use/land cover changes*. Maastricht: RIKS.

Van den Belt, M. (2004). Mediated modelling. *A System Dynamics Approach to*.

Villasante, T., Montañes, M., & Marti, J. (2000). *La investigación social participativa, construyendo ciudadanía/1*. Barcelona: Ediciones El Viejo Topo.

Villasante, T. (2015). Conjuntos de acción y grupos motores para la transformación ambiental. *Política y Sociedad, 52*(2), 387–408.

Voinov, A., & Gaddis, E. J. B. (2008). Lessons for successful participatory watershed modeling: a perspective from modeling practitioners. *Ecological Modelling, 216*(2), 197–207.

Volkery, A., & Ribeiro, T. (2007). Prospective environmental analysis of land use development in Europe: from participatory scenarios to long-term strategies. In *Amsterdam conference on the human dimensions of global environmental change "Earth system governance: Theories and strategies for sustainability.*

Volkery, A., Ribeiro, T., Henrichs, T., & Hoogeveen, Y. (2008). Your vision or my model? Lessons from participatory land use scenario development on a European scale. *Systemic Practice and Action Research, 21*(6), 459–477.

Walker, B., Holling, C. S., Carpenter, S. R., & Kinzig, A. (2004). Resilience, adaptability and transformability in social-ecological systems. *Ecology and Society, 9*(2), 5. http://www.ecologyandsociety.org/vol9/iss2/art5/.

Walpole, M. J., & Goodwin, H. J. (2001). Local attitudes towards conservation and tourism around Komodo National Park, Indonesia. *Environmental Conservation, 28*(02), 160–166.

Waylen, K. A., Hastings, E. J., Banks, E. A., Holstead, K. L., Irvine, R. J., & Blackstock, K. L. (2014). The need to disentangle key concepts from ecosystem-Approach jargon. *Conservation Biology, 28*(5), 1215–1224.

Weihrich, H. (1982). The TOWS matrix—a tool for situational analysis. *Long Range Planning, 15*(2), 54–66.

Westley, F., Olsson, P., Folke, C., Homer-Dixon, T., Vredenburg, H., Loorbach, D., et al. (2011). Tipping toward sustainability: emerging pathways of transformation. *Ambio, 40*(7), 762–780. http://dx.doi.org/10.1007/s13280-011-0186-9.

White, R., & Engelen, G. (1997). Cellular automata as the basis of integrated dynamic regional modelling. *Environment and Planning B: Planning and Design, 24*(2), 235–246.

Winder, N. (2003). *Successes and problems when conducting interdisciplinary or transdisciplinary (= integrative) research*. Interdisciplinarity and transdisciplinarity in landscape studies: potential and limitations. Delta Program, Wageningen (pp. 74–90).

Winder, N. (2005). Integrative research as appreciative system. *Systems Research and Behavioral Science, 22*(4), 299–309.

Winder, N. (2006). *TiGrESS project final report, report on behalf of European commission*. Newcastle upon Tyne, UK: Newcastle University.

Winder, N. (2007). Innovation and metastability, a systems model. *Ecology and Society, 12*, 2.

Wolsink, M. (2006). Invalid theory impedes our understanding: a critique on the persistence of the language of NIMBY. *Transactions of the Institute of British Geographers, 31*(1), 85–91.

Woolley, A. W., Chabris, C. F., Pentland, A., Hashmi, N., & Malone, T. W. (2010). Evidence for a collective intelligence factor in the performance of human groups. *Science, 330*(6004), 686–688.

World Agroforestry Center (WAC). (2003). *Evaluation dartboard.* http://www.cglrc.cgiar.org/icraf/toolkit/Evaluation_dartboard.htm.

Yelland, P. M. (2010). An introduction to correspondence analysis. *Math. J, 12*, 1–23. Available from https://pdfs.semanticscholar.org/c6a8/34c7d443c9feea86568cdab0346ba424db44.pdf.

Yuan, Z., Lun, F., He, L., Cao, Z., Min, Q., Bai, Y., et al. (2014). Exploring the state of retention of traditional ecological knowledge (TEK) in a Hani rice terrace village, Southwest China. *Sustainability, 6*(7), 4497–4513.

Further Reading

Alberich, T. (1998). Introducción a los métodos y técnicas de investigación social y la IAP. In *Cuadernos de la Red* (Vol. 5, pp. 31–41). Madrid: Red CIMS.

Barreto Dillon, L. Problem tree analysis (Undated). *Sustainable sanitation and water management.* Available at: http://www.sswm.info/content/problem-tree-analysis.

Basagoiti, R. M., Bru, M. P., & Lorenzana, C. (2001). *La IAP de bolsillo.* Madrid: Edita ACSUR Las Segovias.

Berkes, F. (2009). Evolution of co-management: role of knowledge generation, bridging organizations and social learning. *Journal of Environmental Management, 90*(5), 1692–1702.

Burbano Trimiño, F. A. (2013). *Las migraciones internas durante el franquismo y sus efectos sociales: El caso de Barcelona.*

Carasik, L. (2016). *N dakota pipeline protest is a harbinger of many more.* ALJAZEERA. available at http://www.aljazeera.com/indepth/opinion/2016/11/dakota-pipeline-protest-harbinger-161120150300919.html.

Chambers, R. (1983). *Rural development: Putting the last first.* London: Longmans.

Cohen, S. J. (1997). Scientist–stakeholder collaboration in integrated assessment of climate change: lessons from a case study of Northwest Canada. *Environmental Modeling & Assessment, 2*(4), 281–293.

Index

Printed in the United States
By Bookmasters